Tweeting
LINUX

140 Linux Configuration Commands Explained in 140 Characters or Less

First Edition
Version 1.1

Don R. Crawley
Linux+ and CCNA Security

soundtraining.net
accelerated i.t. training

Seattle, Washington
www.soundtraining.net

Special discounts are available on bulk quantities of soundtraining.net books. For details, contact soundtraining.net, a division of Jonan, Limited, PO Box 48094, Seattle, WA 98148.

Telephone: (206) 988-5858

Email: info@soundtraining.net

Website: www.soundtraining.net

Cover and interior design by Jason Sprenger, Overland Park, Kansas, *www.fourthcup.org*
Back cover photograph: JMC Photography, Seattle, Washington

Reasonable attempts have been made to ensure the accuracy of the information contained in this publication as of the date on which it was written. This publication is distributed in the hope that it will be helpful, but with no guarantees. There are no guarantees made as to the accuracy, reliability, or applicability of this information for any task or purpose whatsoever.

The author recommends that these procedures be used only as a guide to configuration of computers and/or devices in a test environment prior to usage in a production environment. Under no circumstances should these procedures be used in a live, production environment without first being tested in a laboratory environment to determine their suitability, their accuracy, and any security implications.

ISBN: 978-0-9836607-1-2

Also available as an e-book in various formats. Please check availability with your favorite e-book retailer.

soundtraining . net
accelerated i.t. training

PO Box 48094
Seattle, Washington 98148-0094
United States of America

On the web: www.soundtraining.net
On the phone: 206.988.5858
Email: info@soundtraining.net

Dedicated to Janet, Bea, Jon, Andrew, Ellie,
Chloe, Cleo, and Frank.

"Technology, like art, is a soaring exercise of the human imagination."

Daniel Bell, "The Winding Passage", 1980

Preface

The idea for this book came from attending a seminar on leadership with Kevin Eikenberry, *(kevineikenberry.com)* who tweets on leadership *(@KevinEikenberry)* and who put his tweets in a book.

I love playing with computers and especially operating systems. If I can find an example of how to do a particular task, I can always figure out other things to do that are related to that task. That's also the attitude I take in writing books. There are lots of big, thick books with lots of details, there's always the Internet with its myriad websites, forums, and blogs, but there don't seem to be very many resources that are short and concise with lots of examples. This book is designed to get you started. I believe you're perfectly capable of figuring things out on your own, if someone will just give you a starting point, so here it is!

To test things while I was writing it, I used VMWare Workstation 6.5 with CentOS 5.6, CentOS 6, Ubuntu Server 10.04, Ubuntu Desktop 11.04, and Debian 6.0.2.1.

These are commands that I use, that people I know use, or that I discovered while writing the book that I thought were useful or cool. If I've omitted your favorite command, let me know. My email is *don@soundtraining.net.*

The most important thing is to enjoy the process of learning and experimenting, to do good work, and, like a doctor, to do no harm. So, test this stuff in a VM or some other lab before you put it into production, make a backup, and double-check your syntax.

I hope you enjoy using this book as much as I enjoyed putting it together.

Online Companion Resources

Errors and Updates: An errata page is maintained at www.soundtraining.net/bookstore/errata

Videos: Many free videos are available at www.soundtraining.net/videos

Web Page: There is a supporting web page with live links and other resources for this book at www.soundtraining.net/tweeting-linux

Facebook: www.soundtraining.net/facebook

Twitter: www.soundtraining.net/twitter

Blog: www.soundtraining.net/blog

```
root@cf:~                                                    ▢ ▣ ✕
[root@cf ~]# adduser -c "Jonathon Y. Crawley" -G athletes jcrawley
[root@cf ~]# ▯
```

Figure 1.1

In the above screen capture, adduser created a new user with a comment (-c) of Jonathon Y. Crawley, who is a member of the secondary group (-G) athletes, and whose username is jcrawley.

```
root@cf:~                                                    ▢ ▣ ✕
[root@cf ~]# adduser -c "Jonathon Y. Crawley" -G athletes jcrawley
[root@cf ~]# less /etc/passwd
```

Figure 1.2

In the second screen capture, the less command is used to display the contents of /etc/passwd so we can see the newly added user.

```
don:x:500:500:Don R. Crawley:/home/don:/bin/bash
mysql:x:27:27:MySQL Server:/var/lib/mysql:/bin/bash
jcrawley:x:501:502:Jonathon Y. Crawley:/home/jcrawley:/bin/bash
(END)
```

Figure 1.3

In the third screen capture, you see the newly added user at the bottom of /etc/passwd. Note that each of the fields are separated by colons (:). The first field represents the username, the second field indicates by the "x" that shadow passwords are being used, 501 is the user ID (UID), 502 is the group ID (GID), the user's full name (Jonathon Y. Crawley) is the comment field, /home/jcrawley is the user's home directory, and /bin/bash is the user's default shell.

(Continued on page 10)

a

adduser

@soundtraining

adduser: creates a new user
account and UID which can be
found in /etc/passwd

1 day ago

In Debian Linux, the adduser command runs a script which asks for additional user information, as you can see in the fourth screen capture from a system running Debian 6.0.2.1.

```
root@debian:~# adduser elianorh
Adding user `elianorh' ...
Adding new group `elianorh' (1001) ...
Adding new user `elianorh' (1001) with group `elianorh' ...
Creating home directory `/home/elianorh' ...
Copying files from `/etc/skel' ...
Enter new UNIX password:
Retype new UNIX password:
passwd: password updated successfully
Changing the user information for elianorh
Enter the new value, or press ENTER for the default
        Full Name []: Elianor S. Holcomb
        Room Number []: 212
        Work Phone []: (206) 555-1234
        Home Phone []: (206) 555-2345
        Other []:
Is the information correct? [Y/n] y
root@debian:~#
```

Figure 1.4

As a Debian-based distribution, Ubuntu behaves similarly with the adduser command.

See also "useradd", "userdel", and "usermod".

a

adduser

@soundtraining

adduser: creates a new user account and UID which can be found in /etc/passwd

1 day ago

```
root@inst-centos:~                                          _  □  X
[root@inst-centos ~]# alias
alias cp='cp -i'
alias l.='ls -d .* --color=tty'
alias ll='ls -l --color=tty'
alias ls='ls --color=tty'
alias mv='mv -i'
alias rm='rm -i'
alias which='alias | /usr/bin/which --tty-only --read-alias --show-dot --show-ti
lde'
[root@inst-centos ~]# alias etc="cd /etc"
[root@inst-centos ~]# alias
alias cp='cp -i'
alias etc='cd /etc'
alias l.='ls -d .* --color=tty'
alias ll='ls -l --color=tty'
alias ls='ls --color=tty'
alias mv='mv -i'
alias rm='rm -i'
alias which='alias | /usr/bin/which --tty-only --read-alias --show-dot --show-ti
lde'
[root@inst-centos ~]#
```

Figure 2.1

Note in the screen capture how the output from the first use of the alias command shows pre-configured aliases. Then, after using the alias command to create the new alias "etc", the output includes the new alias.

Using alias in this manner creates an alias only for the duration of the current logon session. In order to make it persistent across logons, you must add it to the hidden .bashrc file in your home directory.

See also "unalias".

a

alias

@soundtraining

alias: creates a new alias when used with options or, when used without options lists aliases to standard output (STDOUT)

30 minutes ago

```
root@inst-centos:~
[root@inst-centos ~]# apachectl
Usage: /usr/sbin/httpd [-D name] [-d directory] [-f file]
                       [-C "directive"] [-c "directive"]
                       [-k start|restart|graceful|graceful-stop|stop]
                       [-v] [-V] [-h] [-l] [-L] [-t] [-S]
Options:
  -D name            : define a name for use in <IfDefine name> directives
  -d directory       : specify an alternate initial ServerRoot
  -f file            : specify an alternate ServerConfigFile
  -C "directive"     : process directive before reading config files
  -c "directive"     : process directive after reading config files
  -e level           : show startup errors of level (see LogLevel)
  -E file            : log startup errors to file
  -v                 : show version number
  -V                 : show compile settings
  -h                 : list available command line options (this page)
  -l                 : list compiled in modules
  -L                 : list available configuration directives
  -t -D DUMP_VHOSTS  : show parsed settings (currently only vhost settings)
  -S                 : a synonym for -t -D DUMP_VHOSTS
  -t -D DUMP_MODULES : show all loaded modules
  -M                 : a synonym for -t -D DUMP_MODULES
  -t                 : run syntax check for config files
[root@inst-centos ~]#
```

Figure 3.1

The apachectl command is similar to the httpd command. Many functions are the same, but the command syntax is different.

On Debian systems, the httpd command doesn't appear to be available, so you must use apachectl. On RedHat-based systems, both commands are available.

See also "httpd".

a

apachectl

@soundtraining

apachectl: is a control interface for the Apache HTTP server. Use without options to display a list of options.

26 minutes ago

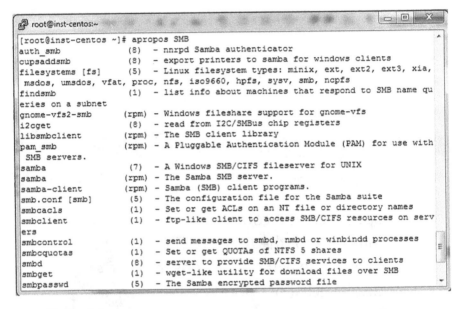

Figure 4.1

In this screen capture, apropos found instances of man pages that include the text string "SMB".

See also "man".

a

apropos
@soundtraining

apropos: searches the descriptions of
man pages for the specified keyword

54 minutes ago

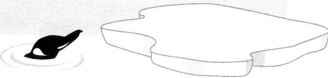

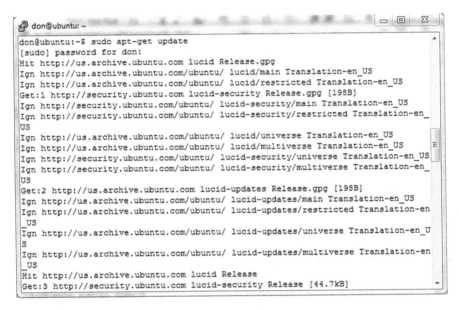

```
don@ubuntu: ~
don@ubuntu:~$ sudo apt-get update
[sudo] password for don:
Hit http://us.archive.ubuntu.com lucid Release.gpg
Ign http://us.archive.ubuntu.com/ubuntu/ lucid/main Translation-en_US
Ign http://us.archive.ubuntu.com/ubuntu/ lucid/restricted Translation-en_US
Get:1 http://security.ubuntu.com lucid-security Release.gpg [198B]
Ign http://security.ubuntu.com/ubuntu/ lucid-security/main Translation-en_US
Ign http://security.ubuntu.com/ubuntu/ lucid-security/restricted Translation-en_
US
Ign http://us.archive.ubuntu.com/ubuntu/ lucid/universe Translation-en_US
Ign http://us.archive.ubuntu.com/ubuntu/ lucid/multiverse Translation-en_US
Ign http://security.ubuntu.com/ubuntu/ lucid-security/universe Translation-en_US
Ign http://security.ubuntu.com/ubuntu/ lucid-security/multiverse Translation-en_
US
Get:2 http://us.archive.ubuntu.com lucid-updates Release.gpg [198B]
Ign http://us.archive.ubuntu.com/ubuntu/ lucid-updates/main Translation-en_US
Ign http://us.archive.ubuntu.com/ubuntu/ lucid-updates/restricted Translation-en
_US
Ign http://us.archive.ubuntu.com/ubuntu/ lucid-updates/universe Translation-en_U
S
Ign http://us.archive.ubuntu.com/ubuntu/ lucid-updates/multiverse Translation-en
_US
Hit http://us.archive.ubuntu.com lucid Release
Get:3 http://security.ubuntu.com lucid-security Release [44.7kB]
```

Figure 5.1

In this screen capture from a computer running Ubuntu server, the apt-get command, used in conjunction with sudo and the "update" option, queries various sources for package updates and displays the results. You could then use apt-get with the "upgrade" option to upgrade the packages.

See the "sudo" page, later in the book, for information about the use of sudo.

See also "aptitude", "rpm" and "yum".

a

apt-get
@soundtraining

apt-get: in Debian-based
distros (such as Ubuntu)
installs and manages
packages (applications)

22 minutes ago

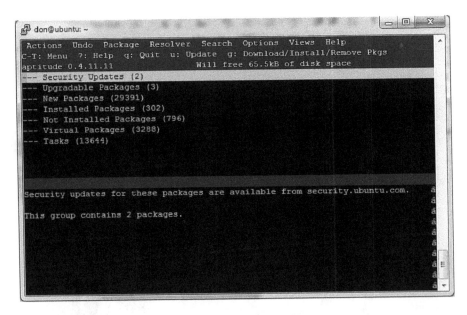

Figure 6.1

You can press "?" for help.

See also "apt-get", "rpm", and "yum".

a

aptitude

@soundtraining

aptitude: in Debian-based distros (such as Ubuntu), an interface for the package manager

1 day ago

```
root@inst-centos:~                                                    _ □ X
[root@inst-centos ~]# awk -F: '$3 == 0 { print $1 " is a superuser!" }' /etc/pas
swd
root is a superuser!
[root@inst-centos ~]#
```

Figure 7.1

In this screen capture:

- awk: is the command invoking the awk pattern scanning and processing language

- -F: tells awk that the field separator is ":"

- $3==0 tells awk to look for a value of "0" in the third field

- Print $1 tells awk to output the first field (in this case, the username)

- " is a superuser!" is simply a text string appended to the output

- /etc/passwd tells awk to scan /etc/passwd

a

awk

@soundtraining

awk: a data extraction and reporting tool which can be used against textual data, both in files and in datastreams.

10 minutes ago

```
root@inst-centos:~                                    □ ▣ X
[root@inst-centos ~]# ll
total 40
-rw-------  1 root root  1406 May 21 11:14 anaconda-ks.cfg
-rw-r--r--  1 root root 25834 Jul 10 07:21 bootmessages.txt
drwxr-xr-x  2 root root  4096 May 22 07:02 Desktop
[root@inst-centos ~]# bzip2 bootmessages.txt
[root@inst-centos ~]# ll
total 20
-rw-------  1 root root 1406 May 21 11:14 anaconda-ks.cfg
-rw-r--r--  1 root root 7205 Jul 10 07:21 bootmessages.txt.bz2
drwxr-xr-x  2 root root 4096 May 22 07:02 Desktop
[root@inst-centos ~]# ▏
```

Figure 8.1

Notice in the screen capture how bzip2 was used to compress the file bootmessages.txt from a file size of 25,834 bytes down to 7,205 bytes.

You can also implement bzip2 compression in conjunction with tar. See the tar page for details.

Uncompress bzip2-compressed files with the command bunzip2 followed by the filename.

See also "gzip" and "zip".

b

bzip2
@soundtraining

bzip2: a compression utility which can be used by itself or in conjunction with tar

2 days ago

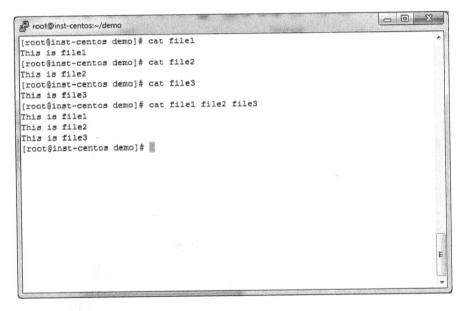

Figure 9.1

In the above example, I used cat to display the contents of three files individually, then combined the three files into one screen output.

See also "less".

C

cat

@soundtraining

cat: short for concatenate, cat allows you to view and modify a file or combine multiple files to standard output

34 minutes ago

```
root@inst-centos:~/demo                                    _  □  X
[root@inst-centos ~]# ls -l
total 24
-rw-------  1 root root 1406 May 21 11:14 anaconda-ks.cfg
-rw-r--r--  1 root root 7205 Jul 10 07:21 bootmessages.txt.bz2
drwxr-xr-x  2 root root 4096 Jul 18 23:54 demo
drwxr-xr-x  2 root root 4096 May 22 07:02 Desktop
[root@inst-centos ~]# cd demo
[root@inst-centos demo]# 
```

Figure 10.1

In this screen capture, I changed to a child directory from my
current directory, therefore I was able to use the relative path
statement "cd demo". If I wanted, however, to change to a directory
that wasn't a child of the current directory, I would have had to use
an absolute path statement such as "cd /etc/httpd/conf". Notice, in
the first example, the absence of forward slashes and the inclusion
of forward slashes in the second example.

cd

@soundtraining

cd: changes directory

27 minutes ago

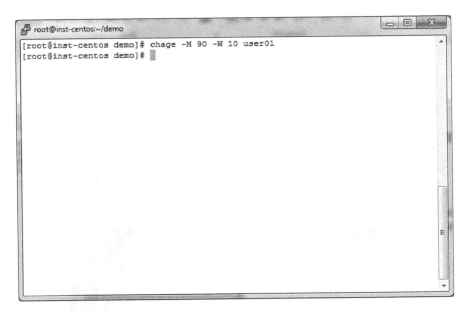

Figure 11.1

In this example, I used chage to require user01 to change passwords every 90 days with a warning to change passwords 10 days prior to expiration.

There are quite a few options for chage. Use "man chage" to see the options.

C

chage

@soundtraining

chage: changes the number of days between password changes and the date of the last password change

1 day ago

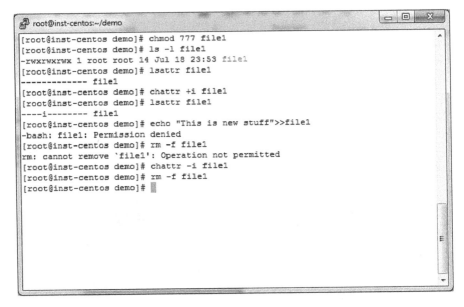

Figure 12.1

In this example, I used chmod to grant full permissions to everyone on file1. Then, I used lsattr to list the attributes of file1. As you can see in the screen capture, there are no attributes set. Next, I used chattr +i to set the immutable attribute on file1. When the immutable attribute is set, the file cannot be removed nor modified in any way. (Although there are many possible attributes, the most commonly-used is "i" for immutable.) The only user who can set the "i" attribute is root. In the second use of lsattr, you can see that the immutable attribute is set. Notice how, even though full permissions are set on file1, it cannot be changed or removed. Once the immutable flag is removed with the chattr –i command, the file can be removed.

As usual, see "man chattr" for more options.

Also see "lsattr".

C

chattr

@soundtraining

chattr: changes file attributes

45 minutes ago

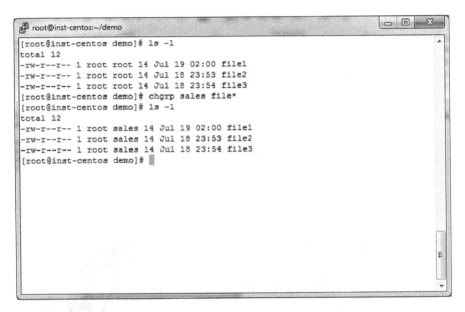

Figure 13.1

Notice that the three files are all owned by the group root in the first example. After the use of chgrp, the group ownership changes to sales.

This produces similar results to chown.

See also "chown".

C

chgrp
@soundtraining

chgrp: changes group ownership

4 hours ago

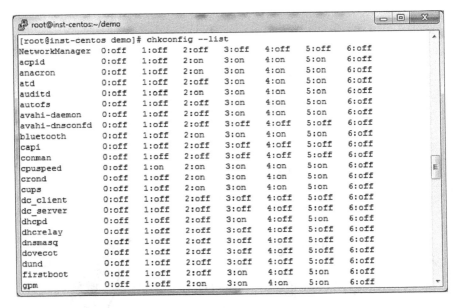

Figure 14.1

In this example, I used chkconfig --list to show each of the daemons (services) and their behavior in each of the runlevels. For example, anacron is off at runlevels 0, 1, and 6, but on at each of the remaining runlevels. If I wanted to view the settings for an individual daemon, I could add its name as an option, for example "chkconfig --list anacron".

(continued on page 38)

C

chkconfig

@soundtraining

chkconfig: displays and modifies run level information, in other words, it allows you to configure which applications start on boot, based on the run level.

30 minutes ago

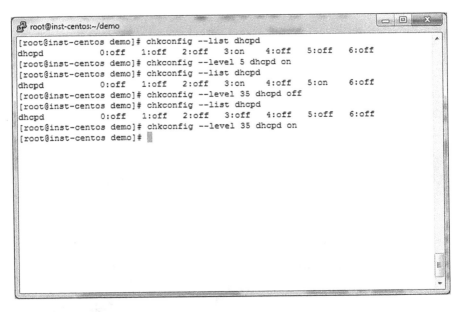

Figure 14.2

In the second screen capture, I first used chkconfig to list the behavior of dhcpd at each of the runlevels. Notice that it's configured to be off at runlevel 5, so in the next command I configured it to be on at runlevel 5. After listing its behaviors again, I then disabled dhcpd in multiple runlevels, and finally re-enabled it in runlevels 3 and 5.

See also "runlevel".

C

chkconfig
@soundtraining

chkconfig: displays and modifies run level information, in other words, it allows you to configure which applications start on boot, based on the run level.

30 minutes ago

```
root@inst-centos:~/demo                                        ▢  X
[root@inst-centos demo]# ls -l
total 12
-rw-r--r-- 1 root root 14 Jul 19 02:00 file1
-rw-r--r-- 1 root root 14 Jul 18 23:53 file2
-rw-r--r-- 1 root root 14 Jul 18 23:54 file3
[root@inst-centos demo]# chmod a+rwx file1
[root@inst-centos demo]# chmod ug+rw file2
[root@inst-centos demo]# chmod o-r file3
[root@inst-centos demo]# ls -l
total 12
-rwxrwxrwx 1 root root 14 Jul 19 02:00 file1
-rw-rw-r-- 1 root root 14 Jul 18 23:53 file2
-rw-r----- 1 root root 14 Jul 18 23:54 file3
[root@inst-centos demo]# ▮
```

Figure 15.1

Notice how chmod a+rwx was used on file1 to add read, write, and
execute (rwx) permissions for all (a). Then, chmod added read and
write (rw) permissions for the user (u) and the group (g) to file2.
Finally, the read permission (r) was removed for other (o) on file3.

(continued on page 42)

C

chmod

@soundtraining

chmod: change or modify file
and directory permissions

44 minutes ago

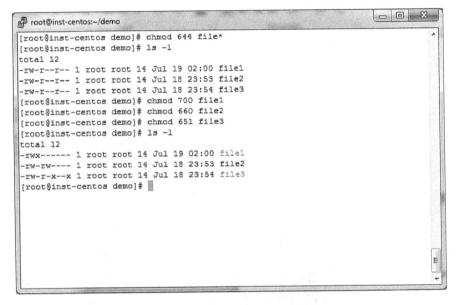

Figure 15.2

Numeric permissions can also be used in which 4 represents read, 2 represents write, and 1 represents execute. They can be used in any combination to produce the desired results.

In the example, 700 (rwx for the user, no permissions for anyone else) was applied to file1, 660 (rw for the user and group, no permission for other) was applied to file2, and 651 was applied to file3, giving rw to the user, rx to the group, and x only to other.

C

chmod

@soundtraining

chmod: change or modify file
and directory permissions

44 minutes ago

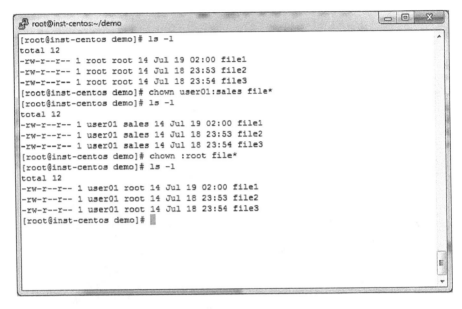

Figure 16.1

Here, chown user01:sales file* changes the ownership of all files whose names start with "file" to the user user01 and the group sales.

Next, chown :root file* changes the group ownership of all files whose names start with "file" to root while maintaining the existing user ownership.

See also "chgrp".

chown

@soundtraining

chown: change user and group ownership of files and directories

2 days ago

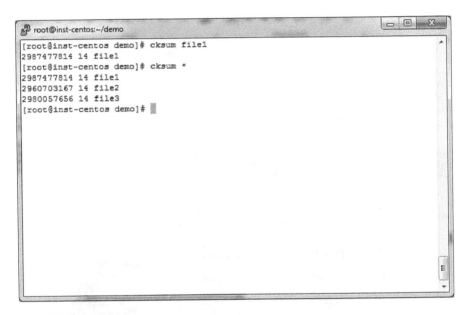

Figure 17.1

In this example, I used cksum on file1 to display its CRC checksum, byte count (14) and filename. Then, I used cksum to display the same information on all the files in the directory.

You can use cksum to validate the integrity of files downloaded via unreliable connections.

C

cksum

@soundtraining

cksum: generates a checksum for a file or a stream of data. It can be used to compare two files to ensure file integrity

54 minutes ago

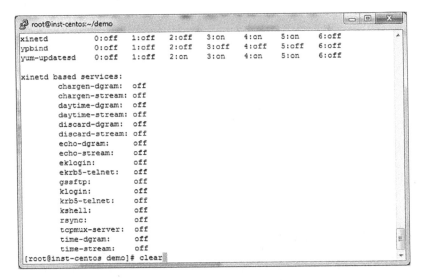

Figure 18.1

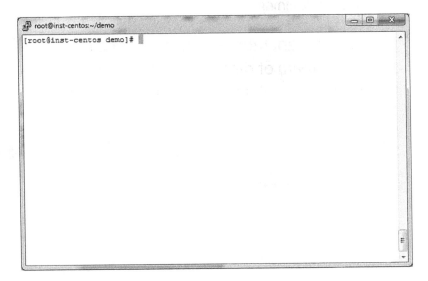

Figure 18.2

There's not a lot to say about "clear", except that it doesn't erase the previous content. You can still scroll back to see it.

C

clear

@soundtraining

clear: clears the screen (similar to cls in Windows)

2 seconds ago

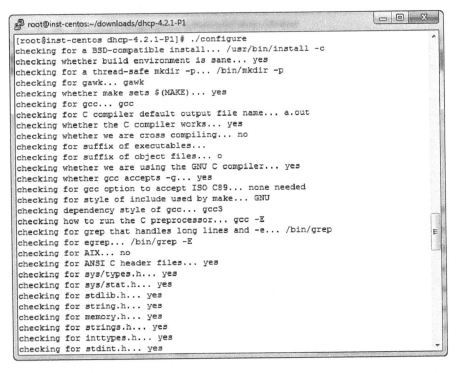

Figure 19.1

In this example, I had previously downloaded and extracted the source code for dhcpd from www.isc.org. After extracting the code, I navigated to the newly-created directory "dhcp-4.2.1-P1" and started the three-step process of installing the software. I first ran ./configure to prepare to run "make" and "make install". (See the "make" page later in the book.)

C

configure
@soundtraining

configure: is a script, usually included in source code downloads, which surveys hardware and checks dependencies before software installs

28 minutes ago

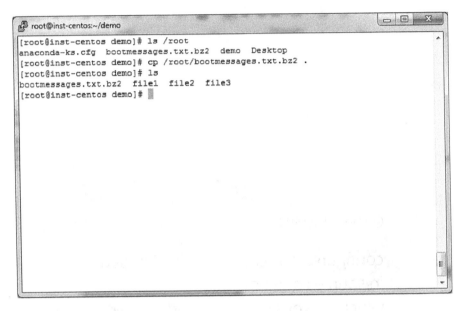

Figure 20.1

In this example, I listed the contents of /root, then used cp to copy bootmessages.txt.bz2 to my working directory (demo). Note the use of the period for the destination, which identifies the current directory as the destination. I could have also explicitly stated a destination with an absolute filename, for example:

```
cp /root/bootmessages.txt.bz2 /root/demo/
bootmessages.txt.bz2
```

See also "mv".

C

cp
@soundtraining

cp: copy a file or directory
15 seconds ago

Figure 21.1

In this example, I opened the crontab file with the command "crontab -e". (You must use the crontab command to open the file, it will not work if you try to open it directly from a text editor.)

Figure 21.2

Here, I created a cron job to delete files in /tmp at 10:45 p.m. on the 1st and 15th of every month.

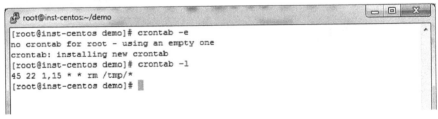

Figure 21.3

Finally, I used the command "crontab -l" to list all of the cron jobs.

C

crontab

@soundtraining

crontab: opens the default text editor such as vim or emacs to edit cron jobs

33 minutes ago

```
root@inst-centos:~
[root@inst-centos ~]# date
Fri Jul 22 07:09:09 PDT 2011
[root@inst-centos ~]# date --set="20110722 08:09"
Fri Jul 22 08:09:00 PDT 2011
[root@inst-centos ~]#
```

Figure 22.1

In the first usage of the command, "date" is simply used to display the current system time.

In the second usage, I set the date and time to Friday, July 22, 2011 at 8:09 a.m.

There are many options available for use with the date command. As usual, see the man or info page. You can also use "date --help" to see options.

date

@soundtraining

date: displays or sets the system date and time

1 day ago

Figure 23.1

In this example, I copied the input partition (if) /dev/sda2 to the output partition (of) /dev/sdb2. Be very careful about the order of input and output statements. Input (from) must go before output (to), otherwise you'll overwrite the original partition with blank data. Oops. (dd is also known as "data destructor".)

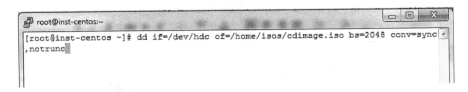

Figure 23.2

In this example, I copied a CD-ROM to an ISO image file named cdimage.iso.

bs=byte size, which is usually expressed in powers of 2, but never less than 512. CDs have a sector-size of 2048 bytes, thus the bs setting in the screen capture.

conv=noerror means continue after read errors

conv=notrunc means do not truncate the output file

conv=sync means if the input block is smaller than the specified input block size, dd pads it to the specified size with null bytes

 **WARNING: USE WITH CAUTION**

dd

@soundtraining

dd: is a bitstream duplicator for copying data. It is especially useful for copying one partition to another

10 minutes ago

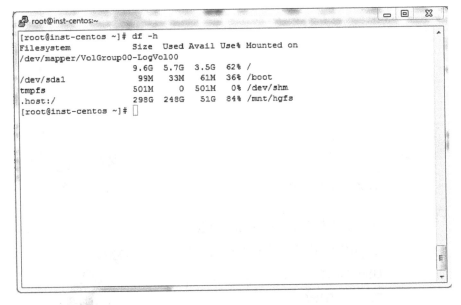

Figure 24.1

This examples shows partition sizes, followed by amount used in megabytes or gigabytes, amount available, amount used as a percentage, and the mount point. (In case you're wondering about the final result (.host:/), that's for the host-guest file system in my VMWare virtual machine from which I got the screen capture.)

d

df

@soundtraining

df: reports the amount of disk usage.
I usually use it with the -h switch to
make its output human-readable

14 minutes ago

Here is the content of the two files used in the example:

file1

```
1   This is some common text at the beginning
2   This is file1
3   This is some common text at the end
4   Dum de dum dum
5   Fee fi fo fum
```

file2

```
1   This is some common text at the beginning
2   This is file2
3   This is some common text at the end
```

```
[root@inst-centos demo]# diff file2 file1
2c2
< This is file2
---
> This is file1
3a4,5
> Dum de dum dum
> Fee fi fo fum
[root@inst-centos demo]#
```

Figure 25.1

In the output of "diff file2 file1":

2c2 means change line two. The "<" symbol refers to the first file, the ">" refers to the second file.

3a4,5 means add lines 4 and 5 in the second file (> (file2) after line 3 in the first file.

d

diff

@soundtraining

diff: finds the differences
between two files

48 minutes ago

```
root@inst-centos:~                                          _ □ X
[root@inst-centos ~]# dig soundtraining.net

; <<>> DiG 9.7.0-P2-RedHat-9.7.0-6.P2.el5_6.3 <<>> soundtraining.net
;; global options: +cmd
;; Got answer:
;; ->>HEADER<<- opcode: QUERY, status: NOERROR, id: 60099
;; flags: qr rd ra; QUERY: 1, ANSWER: 1, AUTHORITY: 2, ADDITIONAL: 0

;; QUESTION SECTION:
;soundtraining.net.              IN      A

;; ANSWER SECTION:
soundtraining.net.      5       IN      A       205.210.189.166

;; AUTHORITY SECTION:
soundtraining.net.      5       IN      NS      ns3.zoneedit.com.
soundtraining.net.      5       IN      NS      ns2.zoneedit.com.

;; Query time: 60 msec
;; SERVER: 192.168.17.2#53(192.168.17.2)
;; WHEN: Mon Jul 25 09:59:57 2011
;; MSG SIZE  rcvd: 99

[root@inst-centos ~]#
```

Figure 26.1

The dig utility, an acronym for "domain information groper" queries
DNS servers for DNS Resource Records such as NS (name server), A
(address or host), SOA (start of authority), TXT (text records, often
used for Sender Policy Framework), PTR (pointer or reverse lookup),
SRV (service), or CNAME (canonical name or aliases).

In this example, I ran dig against the domain name soundtraining.
net and it returned both the host (A) record and the name server
(NS) records. You can specify the type of record in the command,
for example:

```
dig ns soundtraining.net
```

would only return the name server (NS) records.

See also "nslookup" later in this book.

d

dig

@soundtraining

dig: is a DNS utility to query name servers.

22 minutes ago

```
root@inst-centos:~                                                    _  □  X

Linux version 2.6.18-238.19.1.el5 (mockbuild@builder10.centos.org) (gcc version ▲
4.1.2 20080704 (Red Hat 4.1.2-50)) #1 SMP Fri Jul 15 07:31:24 EDT 2011
Command line: ro root=/dev/VolGroup00/LogVol00 rhgb quiet
BIOS-provided physical RAM map:
 BIOS-e820: 0000000000010000 - 000000000009f800 (usable)
 BIOS-e820: 000000000009f800 - 00000000000a0000 (reserved)
 BIOS-e820: 00000000000ca000 - 00000000000cc000 (reserved)
 BIOS-e820: 00000000000dc000 - 00000000000e0000 (reserved)
 BIOS-e820: 00000000000e4000 - 0000000000100000 (reserved)
 BIOS-e820: 0000000000100000 - 000000003fee0000 (usable)
 BIOS-e820: 000000003fee0000 - 000000003feff000 (ACPI data)
 BIOS-e820: 000000003feff000 - 000000003ff00000 (ACPI NVS)
 BIOS-e820: 000000003ff00000 - 0000000040000000 (usable)
 BIOS-e820: 00000000e0000000 - 00000000f0000000 (reserved)
 BIOS-e820: 00000000fec00000 - 00000000fec10000 (reserved)
 BIOS-e820: 00000000fee00000 - 00000000fee01000 (reserved)
 BIOS-e820: 00000000fffe0000 - 0000000100000000 (reserved)
DMI present.
ACPI: RSDP (v002 PTLTD                              ) @ 0x00000000000f6a30
ACPI: XSDT (v001 INTEL   440BX    0x06040000 VMW  0x01324272) @ 0x000000003feed4f
2
ACPI: FADT (v004 INTEL   440BX    0x06040000 PTL  0x000f4240) @ 0x000000003fefee9 ▤
8
:▯                                                                              ▼
```

Figure 27.1

The dmesg command prints bootup messages. It's nearly
impossible to examine the boot messages during startup, so
dmesg allows you to print them to STDOUT or to a file for closer
examination following the boot process.

d

dmesg

@soundtraining

dmesg: prints out boot messages.
You can pipe the output to a
file to simplify examination:
dmesg > bootmessages.txt

2 days ago

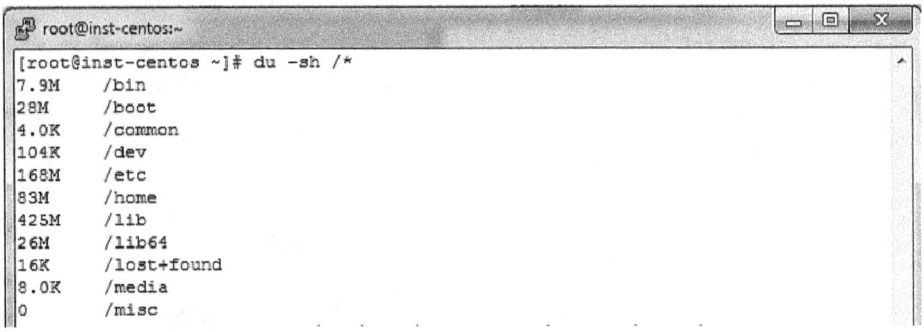

```
root@inst-centos:~

[root@inst-centos ~]# du -sh /*
7.9M    /bin
28M     /boot
4.0K    /common
104K    /dev
168M    /etc
83M     /home
425M    /lib
26M     /lib64
16K     /lost+found
8.0K    /media
0       /misc
```

Figure 28.1

In this example, I used "du -sh /*" to view the size of all directories in a human-readable format.

du

@soundtraining

du: shows disk usage. Commonly used options include -s to summarize and -h to make the output human-readable.

1 day ago

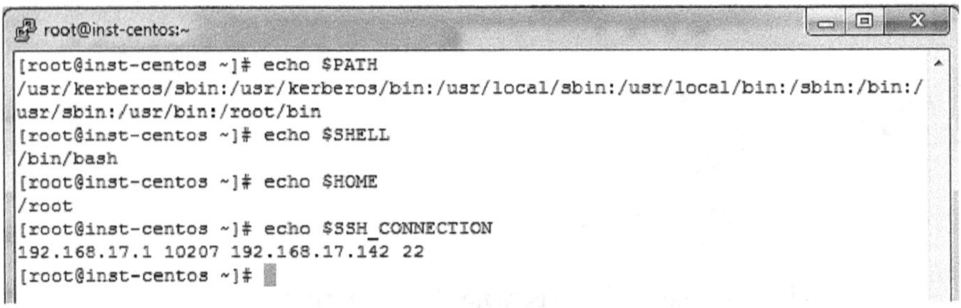

Figure 29.1

Here, I used echo to view a variety of things including the path for the currently logged on user, the current user's default shell and home directory, and information about the currently logged on SSH user.

e

echo
@soundtraining

echo: displays a string of text. It's especially useful in viewing variables such as $HOME, $SHELL, $SSH_CONNECTION, and $PATH

1 day ago

```
root@instructor:~
Disk quotas for user lgaga (uid 501):
  Filesystem                   blocks       soft       hard     inodes       soft       hard
  /dev/sda5                       40          0          0         10          0          0
~
```

Figure 30.1

To enable user and group quotas on a filesystem, you must modify the file /etc/fstab. For specifics about enabling user and group quotas, see the book "The Accidental Administrator: Linux Server Step-by-Step Configuration Guide" or search the Internet.

Also see the following pages in this book: quota, quotacheck, repquota, quotaoff, quotaon, quotastats.

e

edquota

@soundtraining

edquota: edits user quotas

8 minutes ago

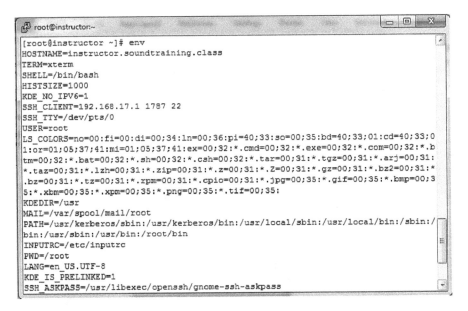

Figure 31.1

The "env" command is a great way to gain an understanding of every aspect of your user environment.

You can also use it to remove variables from your environment with the "-u" (unset) command or to start a program with an empty environment with the "-i" (ignore) command.

e

env

@soundtraining

env: displays the Linux environment variables including path, shell, home directory, login name, and more. It can also be used to edit environment variables.

2 days ago

Figure 32.1

export is not a command in terms of being an executable program. It is an instruction to the shell (probably bash) to make environment variables available to other programs. If you set variables and you want them to be available to other programs, you must export them first.

e

export

@soundtraining

export: defines environment variables. This is non-persistent. To make such variables persistent, modify the user profile in ~/.bashrc or the global profile in /etc/profile or /etc/bashrc

24 minutes ago

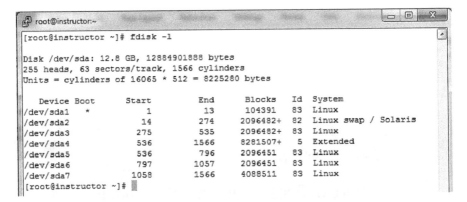

Figure 33.1

It can also be used to set up a new partition, as you can see in the next screen capture.

(Continued on page 80)

WARNING: USE WITH CAUTION

f

fdisk

@soundtraining

fdisk: is a partition table manipulator for Linux. One handy use of fdisk is to display the partition table with the -l option (that's the letter "l", not the number "1").

22 minutes ago

```
root@inst-centos:~

[root@inst-centos ~]# fdisk /dev/sdb
Device contains neither a valid DOS partition table, nor Sun, SGI or OSF disklabel
Building a new DOS disklabel. Changes will remain in memory only,
until you decide to write them. After that, of course, the previous
content won't be recoverable.


The number of cylinders for this disk is set to 1044.
There is nothing wrong with that, but this is larger than 1024,
and could in certain setups cause problems with:
1) software that runs at boot time (e.g., old versions of LILO)
2) booting and partitioning software from other OSs
   (e.g., DOS FDISK, OS/2 FDISK)
Warning: invalid flag 0x0000 of partition table 4 will be corrected by w(rite)

Command (m for help): n
Command action
   e   extended
   p   primary partition (1-4)
p
Partition number (1-4): 1
First cylinder (1-1044, default 1):
Using default value 1
Last cylinder or +size or +sizeM or +sizeK (1-1044, default 1044):
Using default value 1044

Command (m for help): p

Disk /dev/sdb: 8589 MB, 8589934592 bytes
255 heads, 63 sectors/track, 1044 cylinders
Units = cylinders of 16065 * 512 = 8225280 bytes

   Device Boot      Start         End      Blocks   Id  System
/dev/sdb1               1        1044     8385898+  83  Linux

Command (m for help):
```

Figure 33.2

In this screen example, I added a new hard drive to my system
(sdb, which is the second SCSI drive), so I needed to partition it.
I ran the command "fdisk /dev/sdb" to tell the system I wanted to
run fdisk operations on the second SCSI drive. Next, I told it with
the "n" command to create a new partition. The "p" command tells
it to create a primary partition. I chose to make it the first partition
and went with default locations. Finally, I used the "p" command to
print the partition table to STDOUT. Although it's not shown in the
screen capture, I also used the "w" command to write the changes
to the disk.

 WARNING: USE WITH CAUTION

f

fdisk

@soundtraining

fdisk: is a partition table manipulator for Linux. One handy use of fdisk is to display the partition table with the -l option (that's the letter "l", not the number "1").

22 minutes ago

```
root@instructor:~                                                    ☐ ◻ ✕
[root@instructor ~]# file file1
file1: empty
[root@instructor ~]# file file2
file2: ASCII text
[root@instructor ~]# file /bin/hostname
/bin/hostname: ELF 64-bit LSB executable, AMD x86-64, version 1 (SYSV), for GNU/
Linux 2.6.9, dynamically linked (uses shared libs), for GNU/Linux 2.6.9, strippe
d
[root@instructor ~]# █
```

Figure 34.1

In this example, I used the "file" command to display information about an empty file (file1), a text file (file2), and a binary executable file (/bin/hostname).

Although I didn't do it here, try running the "file" command against one of the shell scripts in /etc/init.d.

See also "stat".

f

file

@soundtraining

file: determines and displays
the file type

2 days ago

```
root@instructor:~
[root@instructor ~]# find /etc -name httpd.conf
/etc/httpd/conf/httpd.conf
[root@instructor ~]#
```

Figure 35.1

Some commands are just indispensible and "find" is one of them. Between "find" and "grep", you can usually figure out where anything is on your system.

The "find" command has a ton of options available. In the example, I told the system to search the /etc directory by name (-name) for a file called httpd.conf.

Other example include:

- "find /home −user beatriceh" will find all files under the directory /home owned by user beatriceh
- "find /etc −name *conf" will find all files under the directory /etc that end in conf
- "find /var/log -mtime +30" will find all files under the directory /var/log that were modified more than 30 days ago
- "find . -perm 644" will find files which have read and write permission for their owner, but read-only permission for their group and the world
- "find -iname FindThisFile" will find all files, regardless of case, named "FindThisFile" in the current directory and subdirectories
- "find ~ -empty" will find all empty files in your home directory and subdirectories
- "find . -type f -exec ls -s {} \; | sort -n -r | head -5" will find the five largest files in the current directory and subdirectories

f

find

@soundtraining

find: searches for files in a directory.

3 minutes ago

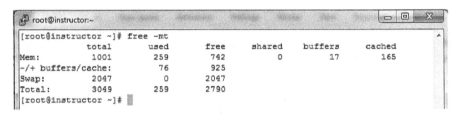

```
root@instructor:~
[root@instructor ~]# free -mt
             total      used      free    shared   buffers    cached
Mem:          1001       259       742         0        17       165
-/+ buffers/cache:        76       925
Swap:         2047         0      2047
Total:        3049       259      2790
[root@instructor ~]#
```

Figure 36.1

In this example, I used the "m" option to display the results in megabytes instead of the default which is kilobytes. I also used the "t" option to display a line with the totals at the bottom.

f

free
@soundtraining

free: displays the amount
of free and used memory in
a system

10 seconds ago

```
root@instructor:~
[root@instructor ~]# fsck /dev/sda5
fsck 1.39 (29-May-2006)
e2fsck 1.39 (29-May-2006)
/home: clean, 136/524288 files, 25582/524112 blocks
[root@instructor ~]#
```

Figure 37.1

```
root@inst-centos:~
[root@inst-centos ~]# fsck -fv /dev/sdb1
fsck 1.39 (29-May-2006)
e2fsck 1.39 (29-May-2006)
Pass 1: Checking inodes, blocks, and sizes
Pass 2: Checking directory structure
Pass 3: Checking directory connectivity
Pass 4: Checking reference counts
Pass 5: Checking group summary information

       15 inodes used (0.00%)
        1 non-contiguous inode (6.7%)
          # of inodes with ind/dind/tind blocks: 0/0/0
    70326 blocks used (3.35%)
        0 bad blocks
        1 large file

        2 regular files
        3 directories
        0 character device files
        0 block device files
        0 fifos
        0 links
        1 symbolic link (1 fast symbolic link)
        0 sockets
--------
        6 files
[root@inst-centos ~]#
```

Figure 37.2

In the first example, I simply ran fsck against /dev/sda5 which had no errors and which fsck reported as clean.

In the second example, I included the options "-fv" to force ("f") checking even though the filesystem was clean and to produce verbose ("v") output.

The Windows equivalent is "chkdsk".

f

fsck

@soundtraining

fsck: is the Linux file system checker.
It can also be used to repair the
file system

54 minutes ago

Figure 38.1

I find it easier to stick with vi(m) when I'm working with Linux systems, even if I have a GUI running, but a lot of people like gedit because it is a lot like Windows Notepad. Use whatever you're most comfortable with and whatever makes your job easier.

See also "vi".

g

gedit
@soundtraining

gedit: is a text editor for the Gnome desktop.

30 minutes ago

```
root@inst-centos:~                                              - □ X
[root@inst-centos ~]# grep -Hrn Skinner /home
/home/don/.bash_history:27:echo "Skinner Pipe Organs">file1
/home/don/file1:1:Skinner Pipe Organs
[root@inst-centos ~]#
```

Figure 39.1

I first became a fan of "grep" when I was searching for a particular text string on a Web server. Although it can sometimes be a little slow, depending on the nature of the search, it will ultimately find what you're looking for.

Other uses include filtering output as shown in the example below.

```
root@inst-centos:~                                              - □ X
[root@inst-centos ~]# ps aux | grep httpd
root      4037  0.0  1.5 259696 15776 ?      Ss   Aug10  0:03 /usr/sbin/httpd
root     24152  0.0  0.0  59132   836 pts/1  S+   09:47  0:00 man httpd
root     24154  0.0  0.1  63876  1060 pts/1  S+   09:47  0:00 sh -c /usr/bin/bzip
2 -c -d /var/cache/man/cat8/httpd.8.bz2 | /usr/bin/less -is
apache   24273  0.0  0.7 259696  8192 ?      S    09:49  0:00 /usr/sbin/httpd
apache   24274  0.0  0.7 259696  8188 ?      S    09:49  0:00 /usr/sbin/httpd
apache   24275  0.0  0.7 259696  8188 ?      S    09:49  0:00 /usr/sbin/httpd
apache   24276  0.0  0.7 259696  8188 ?      S    09:49  0:00 /usr/sbin/httpd
apache   24277  0.0  0.7 259696  8188 ?      S    09:49  0:00 /usr/sbin/httpd
apache   24278  0.0  0.7 259696  8188 ?      S    09:49  0:00 /usr/sbin/httpd
apache   24279  0.0  0.7 259696  8188 ?      S    09:49  0:00 /usr/sbin/httpd
apache   24280  0.0  0.7 259696  8188 ?      S    09:49  0:00 /usr/sbin/httpd
root     29382  0.0  0.0  61212   744 pts/0  R+   12:00  0:00 grep httpd
[root@inst-centos ~]# rpm -qa | grep ssh
openssh-4.3p2-72.el5_6.3
openssh-askpass-4.3p2-72.el5_6.3
openssh-clients-4.3p2-72.el5_6.3
openssh-server-4.3p2-72.el5_6.3
[root@inst-centos ~]#
```

Figure 39.2

It sure beats trying to manually sort through a "ps aux" or an "rpm -qa" command output!

If this is new to you, I simply ran the output of each of the two commands through a grep filter. The "|" symbol pipes the output of one command into the second command.

g

grep
@soundtraining

grep: is one of the most powerful searching tools in Linux. Among other things, it can find strings of text and print them to standard output

1 day ago

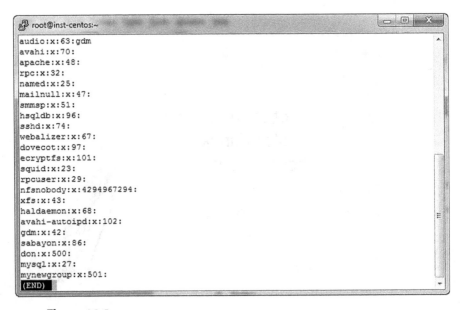

Figure 40.1

Figure 40.2

In this example, I created the group "mynewgroup", then used the command "less /etc/group" to display it. (I could have also piped the output of "less /etc/group" into a grep filter to see just the entry for mynewgroup.

g

groupadd
@soundtraining

groupadd: creates a new usergroup and GID which can be found in /etc/group

12 minutes ago

Figure 41.1

In the first case, I ran the "groups" command with no options and it displayed the group membership for the currently logged on user (root). In the second case, I ran the "groups" command and specified the user andrewc, so it displayed the group membership for user andrewc.

g

groups
@soundtraining

groups: prints the groups a user is in to standard output

27 minutes ago

```
root@inst-centos:~/demo
[root@inst-centos demo]# ls -l
total 40
-rw-r--r-- 1 root    root 25834 Jul 19 15:39 bootmessages.txt
-rw-r--r-- 1 user01 root   121 Jul 22 12:22 file1
-rw-r--r-- 1 user01 root    92 Jul 22 12:22 file2
-rw-r--r-- 1 user01 root    14 Jul 18 23:54 file3
[root@inst-centos demo]# gzip bootmessages.txt
[root@inst-centos demo]# ls -l
total 20
-rw-r--r-- 1 root    root 7198 Jul 19 15:39 bootmessages.txt.gz
-rw-r--r-- 1 user01 root  121 Jul 22 12:22 file1
-rw-r--r-- 1 user01 root   92 Jul 22 12:22 file2
-rw-r--r-- 1 user01 root   14 Jul 18 23:54 file3
[root@inst-centos demo]# gunzip bootmessages.txt.gz
[root@inst-centos demo]# ls -l
total 40
-rw-r--r-- 1 root    root 25834 Jul 19 15:39 bootmessages.txt
-rw-r--r-- 1 user01 root   121 Jul 22 12:22 file1
-rw-r--r-- 1 user01 root    92 Jul 22 12:22 file2
-rw-r--r-- 1 user01 root    14 Jul 18 23:54 file3
[root@inst-centos demo]#
```

Figure 42.1

In this example, I used the "ls –l" command to display the contents of the current directory and the size of each of the files. Notice that bootmessages.txt's filesize is 25,834 bytes.

Next, I ran gzip on bootmessages.txt. Notice in the output of the second "ls –l" command that bootmessages.txt now has a .gz extension and its new filesize is 7,198 bytes.

Finally, I restored bootmessages.txt.gz to bootmessages.txt.

g

gzip
@soundtraining

gzip: compresses and expands files. Use it in conjunction with gunzip

2 seconds ago

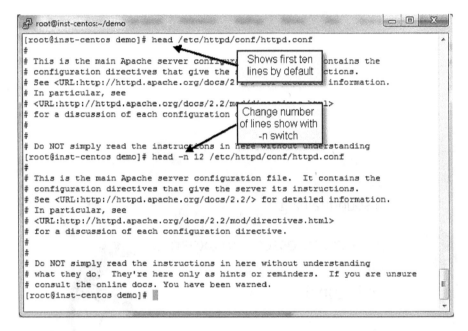

Figure 43.1

The "head" command is useful for taking a quick look at the first few lines of a file.

See also "tail".

h

head

@soundtraining

head: outputs the first part of a file

12 minutes ago

```
root@inst-centos:~/demo
[root@inst-centos demo]# chkconfig --help
chkconfig version 1.3.30.2 - Copyright (C) 1997-2000 Red Hat, Inc.
This may be freely redistributed under the terms of the GNU Public License.

usage:   chkconfig --list [name]
         chkconfig --add <name>
         chkconfig --del <name>
         chkconfig [--level <levels>] <name> <on|off|reset|resetpriorities>
[root@inst-centos demo]#
```

Figure 44.1

Linux "help" files are useful when you just need a quick look at
your command options.

h

help

@soundtraining

help: displays a short help file for commands. The output of –help may include a listing of options, but not lots of detail.

19 minutes ago

```
don@inst-centos:~                                          _  □  X
[don@inst-centos ~]$ history
    1  who
    2  exit
    3  su -
    4  exit
    5  nslookup
    6  dig soa theplatform.com
    7  dig mx soundtraining.net
    8  dig soundtraining.net
    9  host -l soundtraining.class
   10  man host
   11  host -l
   12  host -l soundtraining.class 192.168.0.1
   13  cd /var/named/
   14  su -
   15  exit
   16  smbclient -L -U user01 192.168.0.120
   17  smbclient -U user01 -L 192.168.0.120
   18  mount -t cifs -o username=user01,password=p@ss1234 //192.168.0.120/smbdem
o /mnt/smb02
   19  su -
   20  exit
   21  su -
   22  startx
   23  startx
   24  sudo shutdown -h now
   25  su -
   26  pwd
   27  echo "Skinner Pipe Organs">file1
   28  exit
   29  clear
   30  history
[don@inst-centos ~]$ !26
pwd
/home/don
[don@inst-centos ~]$
```

Figure 45.1

Notice, at the bottom of the screen capture, that you can re-use previously used commands by entering "!" followed by the history number of the command.

h

history

@soundtraining

history: displays recently used commands.

41 minutes ago

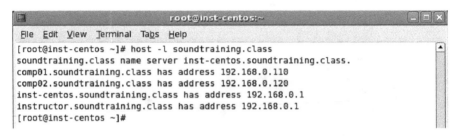

Figure 46.1

In the example, I told "host" to perform a zone transfer. You can also use it with a variety of options such as "-C" to look up the SOA record for a zone or "-t" to specify the type of resource record to look up.

See also "dig" and "nslookup".

h

host

@soundtraining

host: queries a name server

32 minutes ago

Figure 47.1

Simple, but handy in case you're not sure.

h

hostname

@soundtraining

**hostname: displays the
system's hostname**

1 day ago

Figure 48.1

Here, I first used httpd -t to check the syntax of httpd.conf, then I checked the version of Apache with httpd -v, and finally I used httpd -l to output a list of modules compiled in to the Apache server.

See also "apachectl".

h

httpd

@soundtraining

httpd: manipulates the Apache Web server. Options include -v to show the Apache version and -t to check syntax of the httpd.conf file

54 seconds ago

```
andrewc@inst-centos:~                                             [ _ ][ □ ][ X ]
[don@inst-centos ~]$ id
uid=500(don) gid=500(don) groups=500(don)
[don@inst-centos ~]$ su -
Password:
[root@inst-centos ~]# id
uid=0(root) gid=0(root) groups=0(root),1(bin),2(daemon),3(sys),4(adm),6(disk),10
(wheel)
[root@inst-centos ~]# su - andrewc
[andrewc@inst-centos ~]$ id
uid=503(andrewc) gid=507(andrewc) groups=506(research),507(andrewc)
[andrewc@inst-centos ~]$ ▓
```

Figure 49.1

In this example, the user is don who is a member of his own group
(don), then I switched user (su) to root and, as you can see, root is a
member of the groups root, bin, daemon, sys, adm, disk, and wheel,
then I switched user again to andrewc and used the id command to
see that andrewc is a member of his own group (andrewc), plus the
research group.

id

@soundtraining

id: prints the user identity including UID and name, GID and name, and group membership

22 minutes ago

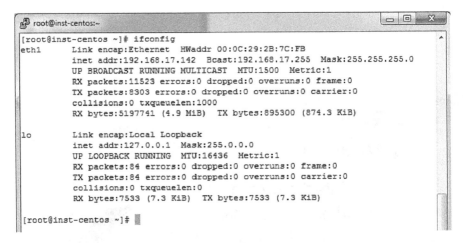

Figure 50.1

Similar to the ipconfig command in Windows, ifconfig displays IP configuration information for each of the interfaces and also includes the hardware address, and a performance summary for each of the interfaces.

ifconfig

@soundtraining

ifconfig: shows interface configuration information including IP address. You can also use it to set an IP address temporarily

22 minutes ago

Figure 51.1

This command shut down interface Ethernet 1.

i

ifdown

@soundtraining

ifdown: disables an interface

10 minutes ago

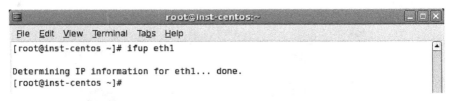

Figure 52.1

This command restarted interface Ethernet 1.

ifup

@soundtraining

ifup: enables an interface

1 day ago

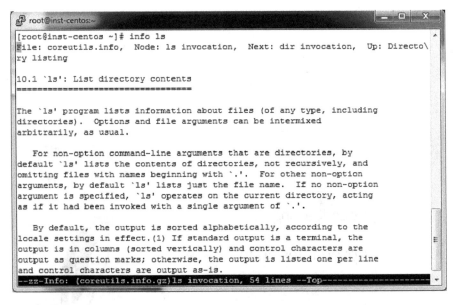

Figure 53.1

The GNU info pages are similar to the man pages, but they're more frequently updated and often easier to read and understand. Still, for some reason, I'm more inclined to use man pages. (Hmmm. I'll have to think about this. I may need to change my ways.)

i

info

@soundtraining

info: show GNU help files

42 minutes ago

```
root@inst-centos:~
[root@inst-centos ~]# init 5
[root@inst-centos ~]#
```

Figure 54.1

In this example, I changed the system from its current runlevel to runlevel 5.

i

init

@soundtraining

init: when used with a runlevel number or letter executes the init script. Can be used to change the current system runlevel

29 minutes ago

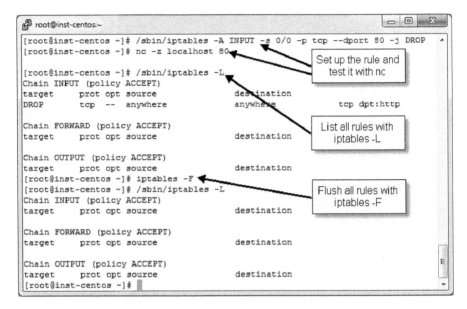

Figure 55.1

iptables has a tremendous amount of complexity. This is a simple example of a very complex command. Here's an explanation:

- The first command appends (-A) the rule to the end of the existing rules, if any exist

- INPUT identifies the source (-s) of the packets (in this case 0/0 means any IP address with any mask)

- Next is the protocol (-p) which is TCP, followed byc the destination port (--dport) which is 80

- -j means jump to the action which, in this case is to DROP the packet

- In other words, this rule will drop packets from any host destined for port 80 (WWW).

iptables

@soundtraining

iptables: is the administration tool for IPv4 packet filtering and network address translation

2 seconds ago

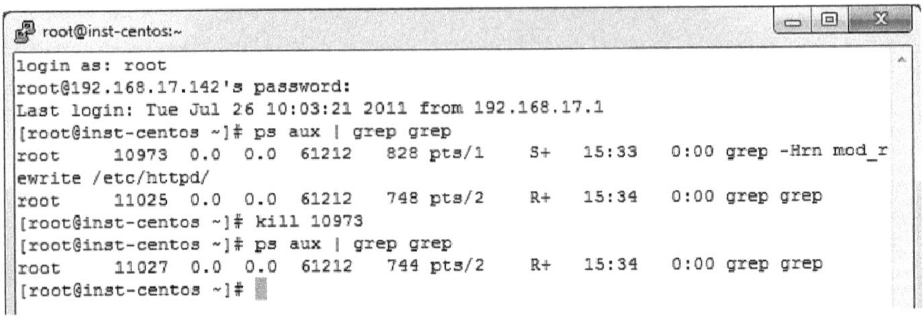

Figure 56.1

Sometimes, a process will hang and you can't seem to stop it. When that happens, the kill command will often do the trick. In this example, I started a command in a different session (grep –Hrn mod_rewrite /etc/httpd/) that I knew would run for a while. Then, I used "ps aux" with a grep filter to view any running process involving grep. In the output of the command, you can see the process in question is #10973. Finally, I used "kill 10973" to stop that particular process.

What kill actually does is send a signal to a process. If no signal is specified, it send a TERM signal, thus stopping the process.

If you're dealing with a process that just won't die, a last resort is "kill -9 XXXXX" (where XXXXX is the process ID).

For more information, see the pages on ps and grep.

k

kill

@soundtraining

kill: terminates a process. In dire situations, you can use the -9 switch to abruptly terminate a process

1 day ago

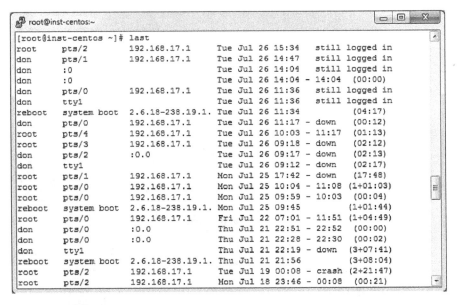

Figure 57.1

This is such a handy command, especially if you work with several administrators or consultants, you can see everyone who has been logged on and when they were logged on. Then, you can go to your system's logs and see what they did.

l

last

@soundtraining

last: shows a listing of the last logged in users

2 minutes ago

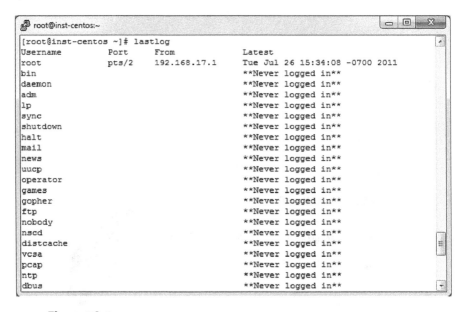

Figure 58.1

This is similar to last, except that it shows all users regardless of whether they actually logged in or not. Use it with grep to check on specific users.

lastlog

@soundtraining

lastlog: reports the most recent login of all users or of a given user

12 minutes ago

```
root@inst-centos:~
[root@inst-centos ~]# less /var/log/messages
Jul 24 04:02:57 inst-centos syslogd 1.4.1: restart.
Jul 24 04:10:20 inst-centos dhclient: DHCPREQUEST on eth1 to 192.168.17.254 port
 67
Jul 24 04:10:20 inst-centos dhclient: DHCPACK from 192.168.17.254
Jul 24 04:10:20 inst-centos dhclient: bound to 192.168.17.142 -- renewal in 761
seconds.
Jul 24 04:23:01 inst-centos dhclient: DHCPREQUEST on eth1 to 192.168.17.254 port
 67
Jul 24 04:23:01 inst-centos dhclient: DHCPACK from 192.168.17.254
Jul 24 04:23:02 inst-centos dhclient: bound to 192.168.17.142 -- renewal in 776
seconds.
Jul 24 04:35:58 inst-centos dhclient: DHCPREQUEST on eth1 to 192.168.17.254 port
 67
Jul 24 04:35:58 inst-centos dhclient: DHCPACK from 192.168.17.254
Jul 24 04:35:58 inst-centos dhclient: bound to 192.168.17.142 -- renewal in 733
seconds.
Jul 24 04:48:10 inst-centos dhclient: DHCPREQUEST on eth1 to 192.168.17.254 port
 67
Jul 24 04:48:11 inst-centos dhclient: DHCPACK from 192.168.17.254
Jul 24 04:48:11 inst-centos dhclient: bound to 192.168.17.142 -- renewal in 687
seconds.
Jul 24 04:59:38 inst-centos dhclient: DHCPREQUEST on eth1 to 192.168.17.254 port
 67
```

Figure 59.1

Unlike "more", "less" allows you to move up and down through a file. In the example, I used "less" to display the contents of /var/log/messages one page at a time. Break out of a "less" display with "q".

Like many of us are fond of saying, "less is so much more than more and more is so much less than less".

Hint: You can search the output of "less" for text strings using "/" followed by the search string.

l

less

@soundtraining

less: displays the contents of a file. Unlike "more", "less" can move backwards and forwards through a file's contents

44 minutes ago

Figure 60.1

Linux supports both hard links and symbolic links. Symbolic links are similar to shortcuts in Windows. Hard links have no equivalent in Windows and are indistinguishable from other files, except that changes in one file are reflected instantly in the other file.

In this screen capture, I created a hard link between file2 and linktofile2. I then used the diff command to compare the files, the echo command to add some text to one, and the cat command to show that the same text was added to both files.

I also used the -s option to create a symbolic link between file1 and file1_symlink.

ln

@soundtraining

ln: makes links between files

25 minutes ago

Figure 61.1

The benefit of "locate" is that it's much faster than, say, find because it queries a database for its results. New files that were added since its last update, however, may not be returned in the query. If you've added a lot of files, be sure to run the "updatedb" command to, as the name implies, update the database.

See also updatedb.

l

locate

@soundtraining

locate: finds files by name.
It uses a database, so
searches tend to be faster
than with "find"

22 seconds ago

Figure 62.1

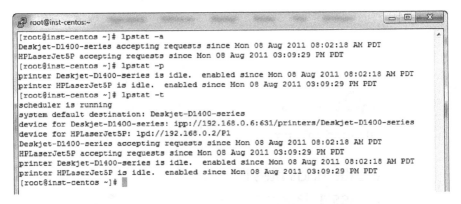

Figure 62.2

The "lpstat" command is handy for displaying jobs in the print queue as you can see in the first example, but it's also helpful for displaying information about printers as you can see in the second example:

- The -a option shows the accepting state of each of the printers

- The -p option shows the status of each of the printers

- The -t option shows all status information

There are many other options. Check the man page, oops, I mean check the info page for details.

l

lpstat

@soundtraining

lpstat: displays printing information. When used with no options, it lists jobs queued by the current user

2 days ago

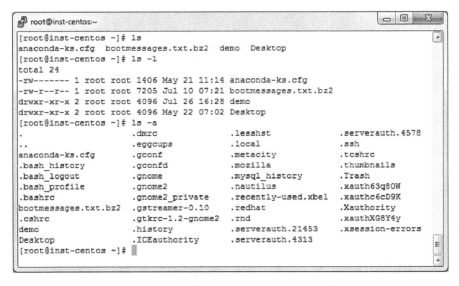

Figure 63.1

In the first example, I simply used ls to view directory contents.
In the second example, I used the -l option to view a long listing
including permissions, ownership, file size, and the date and time of
the creation or last modification.

RedHat-based distros usually alias "ls -l" to "ll".

Another frequently used option with ls is "-a" which shows all files
in a directory, including hidden files (those files whose names start
with a "." such as .bashrc).

l

ls

@soundtraining

ls: lists directory contents. Common options include -l which displays a long listing with more info and -a which shows hidden files

8 minutes ago

```
root@inst-centos:~/demo

[root@inst-centos demo]# lsattr
----i---c---- ./linktofile2
----i---c---- ./file3
------------- ./bootmessages.txt
----i---c---- ./file2
----i---c---- ./file1
[root@inst-centos demo]#
```

Figure 64.1

File attributes can really mess with your head if you're not familiar with them. You can appear to have full control over a file, yet not be able to delete it if, for example, the immutable attribute is set. In this example, file3 has the "i" and "c" attributes set, which makes it immutable (i) and compressed (c). When the "i" attribute is set, the file cannot be deleted, even by the owner or by root. The root user can, however, change the attribute(s) of the file.

For more information, see chattr earlier in this book.

l

lsattr

@soundtraining

lsattr: lists file or directory attributes

3 minutes ago

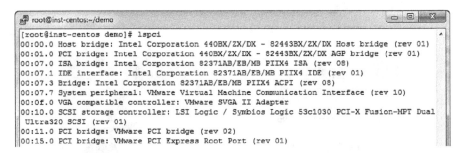

Figure 65.1

In this example, I used lspci to list the PCI devices on a VMWare virtual machine.

l

lspci

@soundtraining

lspci: lists all PCI devices

1 day ago

```
crawleys@kitchen: ~
kitchen
    description: Desktop Computer
    version: 0nx1114RE101NAGAM00
    width: 32 bits
    capabilities: smbios-2.4 dmi-2.4 smp-1.4 smp
    configuration: boot=normal chassis=desktop cpus=1 family=103C_53316J sku=ER8
90AA#ABA uuid=80A8AC11-762D-1110-A1A1-9B1B98B1124D
  *-core
       description: Motherboard
       product: NAGAMI
       vendor: ASUSTek Computer INC.
       physical id: 0
       version: 1.02
       serial: MB-1234567890
     *-firmware
          description: BIOS
          vendor: Phoenix Technologies, LTD
          physical id: 0
          version: 3.11
          date: 09/19/2006
          size: 128KiB
          capacity: 448KiB
          capabilities: pci pnp apm upgrade shadowing cdboot bootselect socketed
:
```

Figure 66.1

The lshw command, available on Debian-based distros such as
Ubuntu, spews forth a lengthy listing of the hardware configuration
of a computer. In this example, I ran it on my family's kitchen
computer, which runs Ubuntu 11.04. The listing goes on much
longer than what I captured.

l

lshw

@soundtraining

lshw: lists hardware

30 seconds ago

```
crawleys@kitchen: ~
crawleys@kitchen:~$ lsusb
Bus 002 Device 005: ID 058f:9360 Alcor Micro Corp. 8-in-1 Media Card Reader
Bus 002 Device 004: ID 0764:0005 Cyber Power System, Inc. Cyber Power UPS
Bus 002 Device 003: ID 03f0:7904 Hewlett-Packard
Bus 002 Device 002: ID 045e:00f9 Microsoft Corp. Wireless Desktop Receiver 3.1
Bus 002 Device 001: ID 1d6b:0001 Linux Foundation 1.1 root hub
Bus 001 Device 005: ID 0424:2504 Standard Microsystems Corp. USB 2.0 Hub
Bus 001 Device 001: ID 1d6b:0002 Linux Foundation 2.0 root hub
crawleys@kitchen:~$
```

Figure 67.1

Another command unique to Debian-based systems, lsusb displays
USB devices on your computer.

l

lsusb

@soundtraining

lsusb: lists USB devices

27 seconds ago

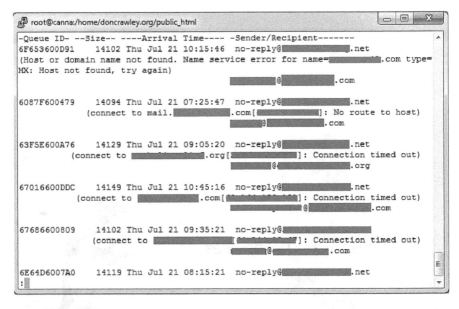

Figure 68.1

In this example, you can see a large number of failed emails that will eventually be deleted from the mail queue. I blacked out addresses to protect user privacy.

m

mailq
@soundtraining

mailq: displays the mail queue

12 minutes ago

```
root@inst-centos:~/downloads/dhcp-4.2.1-P1
[root@inst-centos dhcp-4.2.1-P1]# make
Making all in bind
make[1]: Entering directory `/root/downloads/dhcp-4.2.1-P1/bind'
```

Figure 69.1

```
root@inst-centos:~/downloads/dhcp-4.2.1-P1
[root@inst-centos dhcp-4.2.1-P1]# make install
Making install in bind
make[1]: Entering directory `/root/downloads/dhcp-4.2.1-P1/bind'
make[1]: Nothing to be done for `install'.
make[1]: Leaving directory `/root/downloads/dhcp-4.2.1-P1/bind'
Making install in includes
make[1]: Entering directory `/root/downloads/dhcp-4.2.1-P1/includes'
make[2]: Entering directory `/root/downloads/dhcp-4.2.1-P1/includes'
make[2]: Nothing to be done for `install-exec-am'.
test -z "/usr/include" || /bin/mkdir -p "/usr/include"
 /bin/sh /root/downloads/dhcp-4.2.1-P1/install-sh -c -m 644 'omapip/alloc.h' '/u
```

Figure 69.2

Installing software from source code can be intimidating for new Linux/Unix users, but it's often only a few steps.

- First, you download the tarball (a .tgz file) from a website or an FTP server

- Then you unpack it (usually "tar -xvf [filename].tgz")

- cd to the directory created when you unpacked the tarball

- Run "./configure" (see "configure" earlier in the book)

- Run "make" (as seen above)

- Run "make install" (also as seen above)

In this example, I downloaded the source code for dhcpd from isc.org to install a DHCP server from source. If you've never compiled an application from source code, the ISC DHCP server is a great way to get started (on a test machine, of course). Visit www.isc.org.

See also "configure".

m

make

@soundtraining

make: is used to compile an entire program or pieces of a program

20 minutes ago

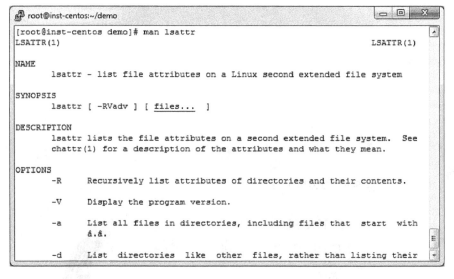

Figure 70.1

The man pages are the tradition Unix programmer's manual documentation pages. They're complete, but often difficult to read and lacking in examples.

See also "help" and "info".

m

man

@soundtraining

man: shows the man pages (the documentation pages)

14 minutes ago

```
root@inst-centos:~
[root@inst-centos ~]# ls
anaconda-ks.cfg  Desktop
[root@inst-centos ~]# mkdir demo
[root@inst-centos ~]# ls
anaconda-ks.cfg  demo  Desktop
[root@inst-centos ~]#
```

Figure 71.1

There's not a lot to say about mkdir.

See also "rmdir".

m

mkdir

@soundtraining

mkdir: creates a new directory

32 seconds ago

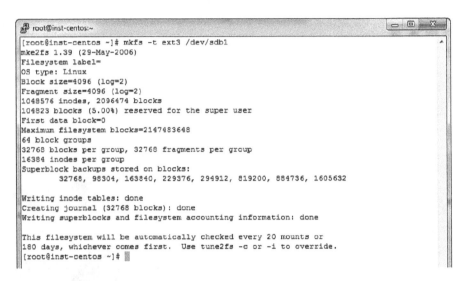

Figure 72.1

The mkfs command makes a new file system. Be careful to ensure you're specifying the partition you think. Otherwise, you could end up wiping out an entire partition irrecoverably. This is one of those commands where it's a good idea to work with a partner to check each other's work before you press the enter key. I'm serious.

 WARNING: USE WITH CAUTION

m

mkfs

@soundtraining

mkfs: builds a Linux file system on a device, usually a hard disk partition

1 day ago

```
root@inst-centos:~                                          ⬒ ▣ ✕
[root@inst-centos ~]# mount 192.168.17.142:/common /mnt/remote
[root@inst-centos ~]# ls /mnt/remote
donsfile  file1  file2  file3  just_go_ahead_and_click_here
[root@inst-centos ~]# umount /mnt/remote
[root@inst-centos ~]# ls /mnt/remote
[root@inst-centos ~]# ▯
```

Figure 73.1

Everything in Linux is based on the file system. If you want to gain access to a resource, say a remote file system, it must first be mounted to your local file system. In this example, I mounted a directory called /common on a computer at 192.168.17.142 and made it available locally through the pre-existing directory /mnt/remote.

You can remove mount points with the command umount.

m

mount

@soundtraining

mount: mounts a file system.
It is often used to gain access
to remote file systems or
removable media such as
a DVD

22 seconds ago

```
root@inst-centos:~
                        My traceroute  [v0.71]
inst-centos.soundtraining.class (0.0.0.0)              Fri Jul 29 17:18:38 2011
Keys:  Help   Display mode   Restart statistics   Order of fields   quit
                                        Packets                Pings
 Host                                   Loss%  Last   Avg  Best  Wrst StDev
 1. 192.168.17.2                        0.0%   2.1    1.4   0.5   3.1   0.5
 2. stackoverflow.com                   1.8% 205.0  162.7  88.0 515.4  78.6
```

Figure 74.1

The mtr command in Linux reminds me of pathping in Windows. It's pretty handy when you're troubleshooting across multiple IP subnets.

It's not installed by default on Debian systems, but you can use "apt-get" to install it.

See also "ping" and "traceroute".

m

mtr

@soundtraining

mtr: combines the functionality of the traceroute and ping programs in a single network diagnostic tool

54 minutes ago

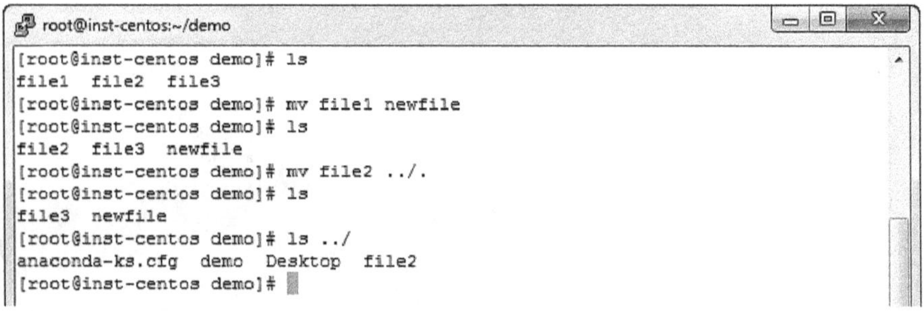

Figure 75.1

Whether you want to actually move a file or directory or simply rename it, you use the mv command to do it. When you want to rename a file, think of it as moving that file to its new name and location, even if it's in the same location as before.

See also "cp".

m

mv

@soundtraining

mv: moves a file or directory from one location to another. Files and directories are renamed by moving them to a new name or location

3 minutes ago

```
root@inst-centos:~                                           ⬓ ⬓ ✕
[root@inst-centos ~]# named-checkconf
[root@inst-centos ~]# named-checkconf -v
9.7.0-P2-RedHat-9.7.0-6.P2.el5_6.3
[root@inst-centos ~]# named-checkconf -z
zone localhost.localdomain/IN: loaded serial 0
zone localhost/IN: loaded serial 0
zone 1.0.0.0.0.0.0.0.0.0.0.0.0.0.0.0.0.0.0.0.0.0.0.0.0.0.0.0.0.0.0.0.ip6.arpa/IN
: loaded serial 0
zone 1.0.0.127.in-addr.arpa/IN: loaded serial 0
zone 0.in-addr.arpa/IN: loaded serial 0
[root@inst-centos ~]#
```

Figure 76.1

When you make changes in your BIND configuration file, it's a good
idea to check the configuration with named-checkconf before
trying to start the server. There are several options available,
including -v to display the version of named-checkconf program
and -z to perform a test load of all the master zones in named.conf.

See also "named-checkzone".

n

named-checkconf

@soundtraining

named-checkconf: checks
the syntax of a BIND DNS
configuration file

1 day ago

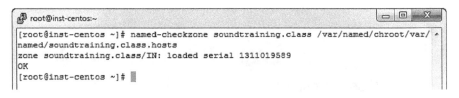

Figure 77.1

This is a very helpful tool for checking a zone file before loading it into a server.

See also "named-checkconf".

n

named-checkzone

@soundtraining

named-checkzone: checks the syntax and integrity of a BIND DNS zone file

30 minutes ago

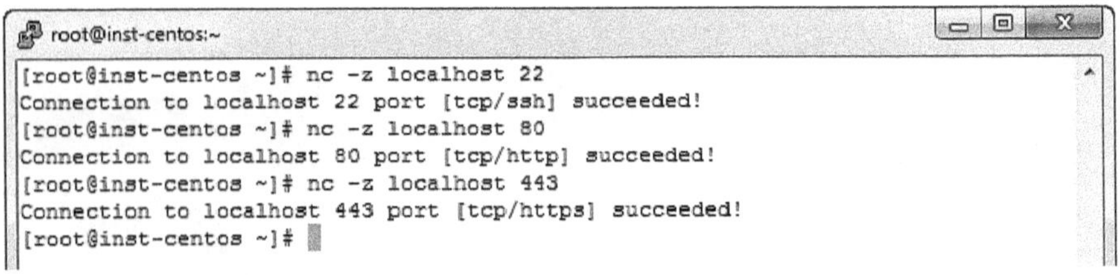

```
root@inst-centos:~
[root@inst-centos ~]# nc -z localhost 22
Connection to localhost 22 port [tcp/ssh] succeeded!
[root@inst-centos ~]# nc -z localhost 80
Connection to localhost 80 port [tcp/http] succeeded!
[root@inst-centos ~]# nc -z localhost 443
Connection to localhost 443 port [tcp/https] succeeded!
[root@inst-centos ~]#
```

Figure 78.1

This is a great tool to make sure you don't have any unexpected ports open on your server.

n

nc

@soundtraining

nc: is netcat, a tool you can use to check for listening connections

2 days ago

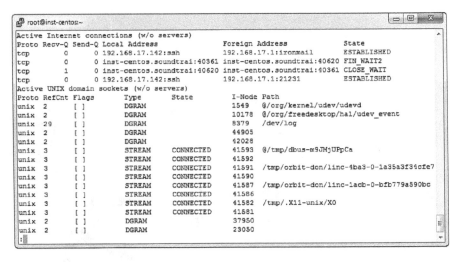

Figure 79.1

Similar to nc, netstat is a great way of surveying what types of connections your server might accept.

n

netstat

@soundtraining

netstat: displays network connections, routing tables, interface statistics, masquerade connections, and multicast memberships

19 minutes ago

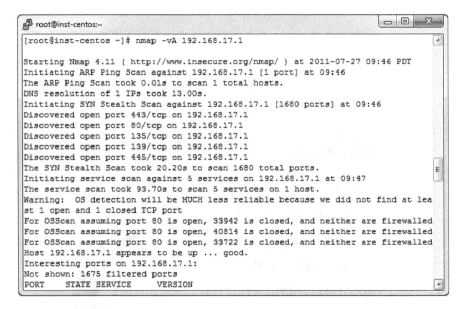

Figure 80.1

This is the original port scanner which is useful for penetration testing.

n

nmap

@soundtraining

nmap: is a network exploration and security analysis tool, including a port scanner

7 minutes ago

Figure 81.1

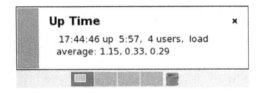

Figure 81.2

The second example is the pop-up that displays near the system tray.

You can use notify-send to display popup messages in the system tray. I heard of a coder who used it to remind himself to stand up and stretch every few hours!

n

notify-send

@soundtraining

notify-send: is used in a GUI to display messages near the system tray

2 days ago

```
root@inst-centos:~

[root@inst-centos ~]# nslookup
> set type=soa
> soundtraining.class
Server:        192.168.0.1
Address:       192.168.0.1#53

soundtraining.class
        origin = inst-centos.soundtraining.class
        mail addr = hostmaster.soundtraining.net
        serial = 1311019589
        refresh = 10800
        retry = 3600
        expire = 604800
        minimum = 38400
> set type=a
> comp01
Server:        192.168.0.1
Address:       192.168.0.1#53

Name:   comp01.soundtraining.class
Address: 192.168.0.110
>
```

Figure 82.1

The nslookup utility queries name servers, similarly to dig, to display information from the zone files. Unlike dig, nslookup includes an interactive mode as seen in the above example.

See also "dig" and "host" earlier in this book.

nslookup

@soundtraining

nslookup: queries DNS servers

52 seconds ago

Figure 83.1

Perhaps the most common use of openssl is to generate a certificate signing request (CSR) to secure a website. Here's an explanation of the syntax in this example:

- Req means I'm creating an X.509 certificate signing request

- -out specifies the output, in this case a file named CSR.csr

- -new generates a new CSR

- -newkey rsa:2048 generates a new private key of type RSA and 2048 bit key length

- -nodes tells openssl not to configure a passphrase to use with the key. It is less secure, of course, than configuring a passphrase, but doesn't require you to enter the passphrase each time you start (or restart) the webserver.

- -keyout specifies the name of the private key, in this case privateKey.key

O

openssl

@soundtraining

openssl: is a command-line tool used for things like generating a certificate-signing request and private key to setup a secure Web server

1 day ago

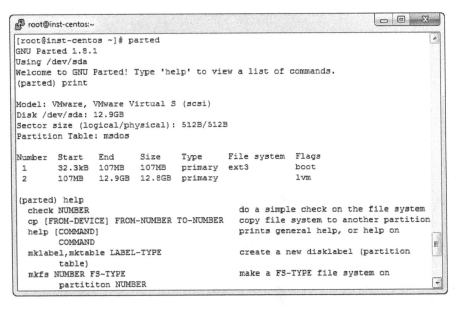

Figure 84.1

GNU parted is a program for creating, destroying, resizing, checking, and copying hard drive partitions, and the file systems on them.

In this example, I simply used the parted command print to display partitions on my system.

See also "fdisk" and "mkfs".

 **WARNING: USE WITH CAUTION**

p

parted

@soundtraining

parted: is a disk partitioning
and resizing program

17 minutes ago

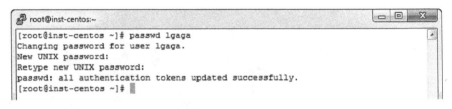

Figure 85.1

In this example, I reset the password for user lgaga. If you enter passwd with no options, it resets the password for yourself.

Linux includes password complexity checking. Non-root users will not be able to set passwords based on dictionary words.

p

passwd
@soundtraining

passwd: resets the specified user's password or, if no user is specified, resets the current user's password

33 minutes ago

Figure 86.1

I wasn't familiar with this command until I started working on this book. It's a handy way to get the process ID of a daemon without having to use "ps aux" with grep.

See also "ps".

p

pidof

@soundtraining

pidof: show the process ID (PID) of running processes when listed by name

9 minutes ago

```
root@inst-centos:~
[root@inst-centos ~]# ping 192.168.17.1
PING 192.168.17.1 (192.168.17.1) 56(84) bytes of data.
64 bytes from 192.168.17.1: icmp_seq=1 ttl=128 time=1.17 ms
64 bytes from 192.168.17.1: icmp_seq=2 ttl=128 time=0.754 ms
64 bytes from 192.168.17.1: icmp_seq=3 ttl=128 time=0.768 ms
64 bytes from 192.168.17.1: icmp_seq=4 ttl=128 time=0.657 ms
64 bytes from 192.168.17.1: icmp_seq=5 ttl=128 time=0.739 ms
64 bytes from 192.168.17.1: icmp_seq=6 ttl=128 time=0.762 ms
64 bytes from 192.168.17.1: icmp_seq=7 ttl=128 time=0.814 ms

--- 192.168.17.1 ping statistics ---
7 packets transmitted, 7 received, 0% packet loss, time 6002ms
rtt min/avg/max/mdev = 0.657/0.809/1.175/0.159 ms
[root@inst-centos ~]#
```

Figure 87.1

Thanks to the late Mike Muuse for inventing the ping utility.
No, contrary to what you may have heard, it does not stand for
"packet internet groper". You can read more about ping by visiting
http://ftp.arl.mil/mike/ping.html.

See also "mtr" and "traceroute".

p

ping
@soundtraining

ping: tests network connectivity
by sending an ICMP ECHO_
REQUEST packet to a specified
host and measuring the
response time

2 seconds ago

```
root@inst-centos:~                                                     [ – ] [ □ ] [ X ]
[root@inst-centos ~]# ps aux
USER       PID %CPU %MEM    VSZ   RSS TTY      STAT START   TIME COMMAND
root         1  0.0  0.0  10372   692 ?        Ss   05:00   0:01 init [5]
root         2  0.0  0.0      0     0 ?        S<   05:00   0:00 [migration/0]
root         3  0.0  0.0      0     0 ?        SN   05:00   0:00 [ksoftirqd/0]
root         4  0.1  0.0      0     0 ?        S<   05:00   0:19 [events/0]
root         5  0.0  0.0      0     0 ?        S<   05:00   0:00 [khelper]
root        14  0.0  0.0      0     0 ?        S<   05:00   0:00 [kthread]
root        18  0.0  0.0      0     0 ?        S<   05:00   0:00 [kblockd/0]
root        19  0.0  0.0      0     0 ?        S<   05:00   0:00 [kacpid]
root       204  0.0  0.0      0     0 ?        S<   05:00   0:00 [cqueue/0]
root       207  0.0  0.0      0     0 ?        S<   05:00   0:00 [khubd]
root       209  0.0  0.0      0     0 ?        S<   05:00   0:00 [kseriod]
root       274  0.0  0.0      0     0 ?        S    05:00   0:00 [khungtaskd]
root       275  0.0  0.0      0     0 ?        S    05:00   0:00 [pdflush]
root       276  0.0  0.0      0     0 ?        S    05:00   0:03 [pdflush]
root       277  0.0  0.0      0     0 ?        S<   05:00   0:00 [kswapd0]
root       278  0.0  0.0      0     0 ?        S<   05:00   0:00 [aio/0]
root       484  0.0  0.0      0     0 ?        S<   05:00   0:00 [kpsmoused]
root       515  0.0  0.0      0     0 ?        S<   05:00   0:00 [mpt_poll_0]
root       516  0.0  0.0      0     0 ?        S<   05:00   0:00 [mpt/0]
root       517  0.0  0.0      0     0 ?        S<   05:00   0:00 [scsi_eh_0]
root       520  0.0  0.0      0     0 ?        S<   05:00   0:00 [ata/0]
root       521  0.0  0.0      0     0 ?        S<   05:00   0:00 [ata_aux]
```

Figure 88.1

The ps utility is one of the most frequently used troubleshooting tools in Linux/Unix. As mentioned in the description, it shows a snapshot of running processes. To get a real-time constantly updating view of running processes, use "top".

Also see "pidof" and "top".

p

ps
@soundtraining

ps: displays a snapshot of
running processes. It is often
used with the options "aux"

17 minutes ago

Figure 89.1

It's easy to get lost and forget where you are. The pwd utility solves that problem.

pwd

@soundtraining

pwd: prints the working (current) directory to STDOUT, usually your terminal

1 day ago

```
root@inst-centos:/users
[root@inst-centos users]# quota lgaga
Disk quotas for user lgaga (uid 501):
    Filesystem blocks   quota   limit   grace   files   quota   limit   grace
    /dev/sdb1       4     750    1000                 1       0       0
[root@inst-centos users]#
```

Figure 90.1

In this example, you can see that user lgaga owns 4 files on /dev/sdb1, she has a soft limit of 750 blocks (1024 bytes per block) and a hard limit of 1000 blocks.

See also "edquota", "quotacheck", "quotaon", "quotaoff", "quotastats", and "repquota".

q

quota

@soundtraining

quota: displays disk usage and limits

29 minutes ago

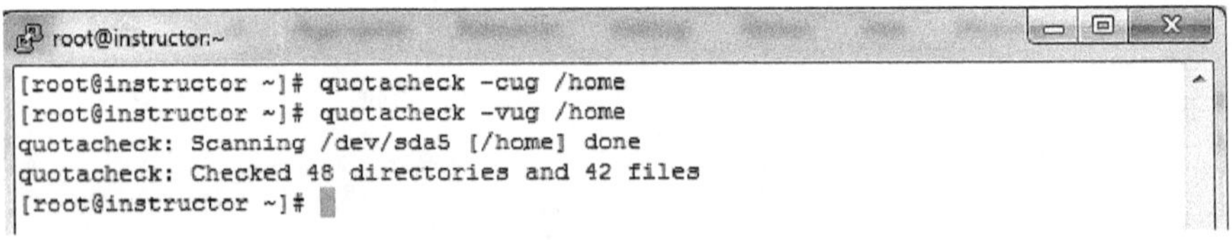

```
root@instructor:~
[root@instructor ~]# quotacheck -cug /home
[root@instructor ~]# quotacheck -vug /home
quotacheck: Scanning /dev/sda5 [/home] done
quotacheck: Checked 48 directories and 42 files
[root@instructor ~]#
```

Figure 91.1

In this example, I first ran quotacheck -cug to prepare quotas on the /home directory, then I ran quotacheck -vug to generate a table of present disk usage.

See also "edquota", "repquota", "quota", "quotaon", "quotaoff", and "quotastats".

q

quotacheck

@soundtraining

quotacheck: scans a filesystem for disk usage. It can also crate, check, and repair quota files.

31 minutes ago

Figure 92.1

In this example, I used the options "g" and "u" to turn off user (u) and group (g) quotas on the file system mounted on /users.

q

quotaoff

@soundtraining

quotaoff: turns filesystem quotas off

18 seconds ago

Figure 93.1

In this example, I used the option "a" to turn on all (a) quotas on all file systems.

```
root@inst-centos:/users
[root@inst-centos users]# quotaon -gu /users
[root@inst-centos users]#
```

Figure 93.2

As you can probably guess, this time I turned on quotas on groups (g) and users (u) on just the file system mounted on /users.

See also "edquota", "repquota", "quota", "quotacheck", and "quotastats".

q

quotaon

@soundtraining

quotaon: turns filesystem
quotas on

3 minutes ago

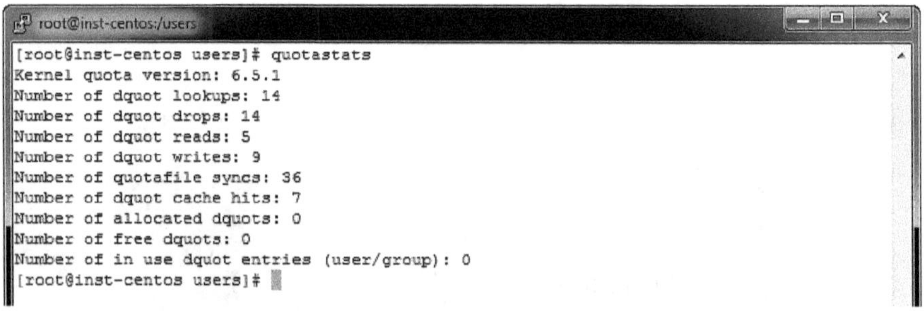

Figure 94.1

This command probes the kernel for quota usage statistics. For most sys admins, more useful output is obtained from repquota.

See also "edquota", "repquota", "quota", "quotacheck", "quotaon", and "quotaoff".

q

quotastats

@soundtraining

quotastats: displays
quota statistics

11 minutes ago

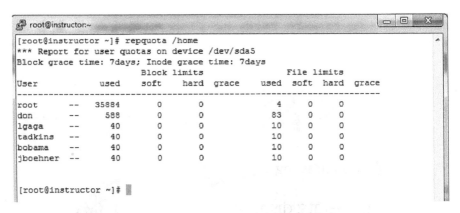

```
root@instructor:~                                              _ □ X
[root@instructor ~]# repquota /home
*** Report for user quotas on device /dev/sda5
Block grace time: 7days; Inode grace time: 7days
                        Block limits              File limits
User            used   soft   hard  grace   used  soft  hard  grace
----------------------------------------------------------------------
root      --   35884     0      0              4     0     0
don       --     588     0      0             83     0     0
lgaga     --      40     0      0             10     0     0
tadkins   --      40     0      0             10     0     0
bobama    --      40     0      0             10     0     0
jboehner  --      40     0      0             10     0     0

[root@instructor ~]#
```

Figure 95.1

The repquota command displays a useful summary of each users disk usage compared to their quota limits.

See also "edquota", "quota", "quotacheck", "quotaon", "quotaoff", and "quotastats".

r

repquota
@soundtraining

repquota: prints a summary
of the disc usage and quotas
for the specified file systems

1 day ago

Figure 96.1

The rm command deletes files irretrievably. Many systems alias rm to rm -i (interactive) to prevent accidental file deletion.

⚠ WARNING: USE WITH CAUTION

rm

@soundtraining

rm: deletes a file or directory

43 minutes ago

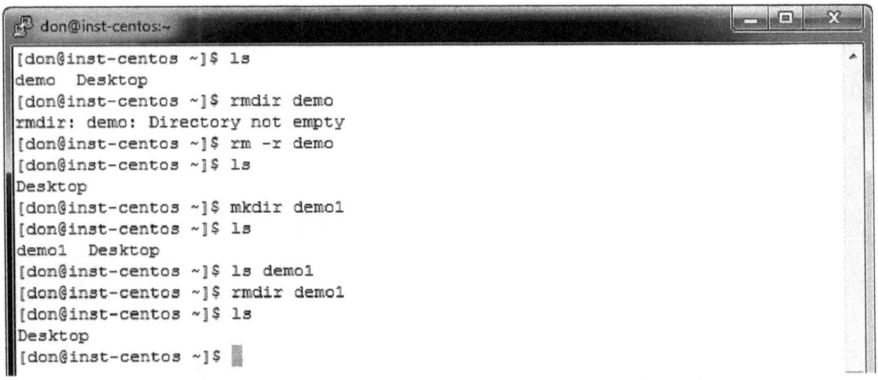

Figure 97.1

This command only works if the directory is empty.

When the directory holds files and I want to delete it, I usually use "rm -rf /directoryname". Just be careful, because you won't be prompted to confirm deletion and there's no retrieving deleted items.

See also "rm".

r

rmdir
@soundtraining

rmdir: deletes a directory, if it
is empty

55 minutes ago

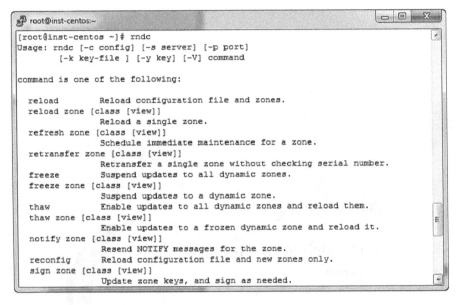

Figure 98.1

Many of the commands you execute with rndc can be executed with various distro tools, but if you install BIND from source, you'll find yourself using rndc a lot.

See also "named-checkconf" and "named-checkzone".

r

rndc

@soundtraining

rndc: is the remote name daemon controller. It allows you to start, stop, and refresh (among other things) the BIND DNS server

2 days ago

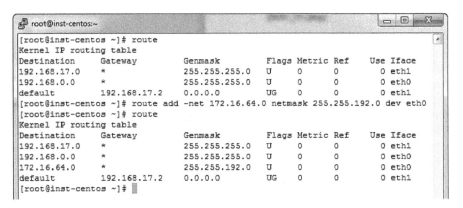

Figure 99.1

As you can, using the route command with no options displays the routing table.

You can also add and delete static routes with the route command. In this example, I added a route to the network 172.16.64.0/18 through interface eth0.

r

route

@soundtraining

route: shows and manipulates
the IP routing table

14 seconds ago

```
root@inst-centos:~
[root@inst-centos ~]# rpm -qa | grep tftp
[root@inst-centos ~]# rpm -ivh /media/CentOS_5.6_Final/CentOS/tftp-server-0.49-2
.el5.centos.x86_64.rpm
Preparing...                ########################################### [100%]
   1:tftp-server           ########################################### [100%]
[root@inst-centos ~]# rpm -qa | grep tftp
tftp-server-0.49-2.el5.centos
[root@inst-centos ~]#
[root@inst-centos ~]# rpm -e tftp-server-0.49-2.el5.centos
[root@inst-centos ~]# rpm -qa | grep tftp
[root@inst-centos ~]#
```

Figure 100.1

The rpm command is useful for installing software, but I usually use yum for installations. I have found rpm very handy, however, for validating software upgrades and showing software version numbers. During a recent security audit, I used rpm to show the auditors that software updates had been installed.

See also "grep".

r

rpm
@soundtraining

rpm: is the command-line environment for the RedHat Package Manager utility

1 day ago

```
root@centos6:~
[root@centos6 ~]# mkdir /test
[root@centos6 ~]# mkdir /mnt/remote
[root@centos6 ~]# mount 192.168.0.9:/common /mnt/remote
[root@centos6 ~]# ls /mnt/remote
donsfile  file1  file2  file3  just_go_ahead_and_click_here
[root@centos6 ~]# ls /test
[root@centos6 ~]# rsync -av /mnt/remote/ /test/
sending incremental file list
./
donsfile
file1
file2
file3
just_go_ahead_and_click_here

sent 343 bytes  received 110 bytes  906.00 bytes/sec
total size is 18  speedup is 0.04
[root@centos6 ~]# ls /test
donsfile  file1  file2  file3  just_go_ahead_and_click_here
[root@centos6 ~]# rsync -av /mnt/remote/ /test/
sending incremental file list
./
donsfile

sent 186 bytes  received 34 bytes  146.67 bytes/sec
total size is 35  speedup is 0.16
[root@centos6 ~]#
```

Figure 101.1

The rsync tool is one of the most valuable tools in your Linux sys admin arsenal. Here's what happened in the example:

- I created a directory called /test and another one called /mnt/remote

- I mounted a remote directory called /common on a host at 192.168.0.9 to the local directory /mnt/remote

- Using "ls /mnt/remote", I observed the contents of /mnt/remote (actually the contents of /common on 192.168.0.9)

- Using "ls /test", I observed that /test is empty

(continued on page 218)

r

rsync

@soundtraining

rsync: allows you to
synchronize files between
two systems

16 minutes ago

```
root@centos6:~                                              _ □ X
[root@centos6 ~]# mkdir /test
[root@centos6 ~]# mkdir /mnt/remote
[root@centos6 ~]# mount 192.168.0.9:/common /mnt/remote
[root@centos6 ~]# ls /mnt/remote
donsfile  file1  file2  file3  just_go_ahead_and_click_here
[root@centos6 ~]# ls /test
[root@centos6 ~]# rsync -av /mnt/remote/ /test/
sending incremental file list
./
donsfile
file1
file2
file3
just_go_ahead_and_click_here

sent 343 bytes  received 110 bytes  906.00 bytes/sec
total size is 18  speedup is 0.04
[root@centos6 ~]# ls /test
donsfile  file1  file2  file3  just_go_ahead_and_click_here
[root@centos6 ~]# rsync -av /mnt/remote/ /test/
sending incremental file list
./
donsfile

sent 186 bytes  received 34 bytes  146.67 bytes/sec
total size is 35  speedup is 0.16
[root@centos6 ~]#
```

Figure 101.2

- I used the command "rsync –av /mnt/remote/ /test/" to synchronize the contents of /mnt/remote with /test

- Using "ls /test", I observed that the contents of /mnt/remote and /test were now synchronized

- Although you can't see it in the screen capture, I modified the file "donsfile" on the remote host

- Once again, I ran "rsync –av /mnt/remote/ /test/" to synchronize the two directories. This time, however, only one file (the file that was changed) was synchronized.

As is often the case, there are many more options available for use with rsync. Read the info page for more information.

r

rsync

@soundtraining

rsync: allows you to synchronize files between two systems

16 minutes ago

Figure 102.1

Linux distros use runlevels to define what starts and stops upon booting. On RedHat-based distros, runlevel 3 is full multi-user mode with no X (graphics) and runlevel 5 is full multi-user mode with X. In this example, the last runlevel used was 3 and the current runlevel is 5.

r

runlevel

@soundtraining

runlevel: displays the current
and previous runlevels

19 minutes ago

```
[don@centos6 ~]$ scp 192.168.0.10:/common/file2 192.168.0.9:/home/don
don@192.168.0.10's password:
The authenticity of host '192.168.0.9 (192.168.0.9)' can't be established.
RSA key fingerprint is 03:1e:86:68:da:a3:fe:e1:6e:61:57:4f:52:2e:11:8e.
Are you sure you want to continue connecting (yes/no)? yes
Warning: Permanently added '192.168.0.9' (RSA) to the list of known hosts.
don@192.168.0.9's password:
file2                                          100%    0     0.0KB/s   00:00
Connection to 192.168.0.10 closed.
[don@centos6 ~]$
```

Figure 103.1

Here, I securely copied file2 from the /common directory on the host at 192.168.0.10 to /home/don on the host at 192.168.0.9.

Notice that SCP first asked for the user's password on the source computer, it then validated the authenticity of the destination computer, asked the user's password on the destination computer, and then transferred the file.

It is certainly possible to transfer a file between a local and a remote computer as well as two remote computers.

You can also use hostnames instead of IP addresses, assuming you have some sort of names resolution enabled.

S

scp
@soundtraining

scp: is secure copy. Similar to rcp, it copies files between hosts on a network. Unlike rcp, it does so securely using SSH

31 seconds ago

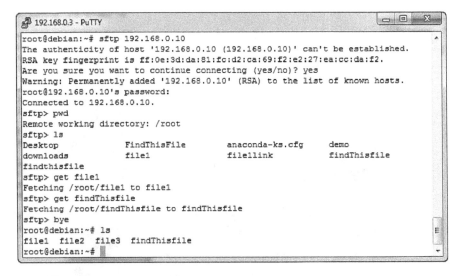

Figure 104.1

In the first screen capture, I logged on as my current user (root), showed the remote working directory (pwd), listed its contents (ls), and downloaded (get) file1.

(Continued on page 226)

S

sftp
@soundtraining

sftp: is the secure file transfer program. Unlike traditional FTP, SFTP provides an encrypted file transfer environment

1 day ago

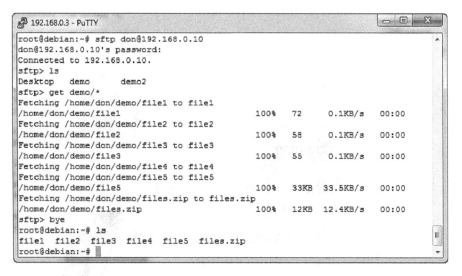

Figure 104.2

In the second example, I logged on as remote user "don" and downloaded the contents of the demo directory (get demo/*).

S

sftp

@soundtraining

sftp: is the secure file transfer program. Unlike traditional FTP, SFTP provides an encrypted file transfer environment

1 day ago

```
root@inst-centos:~
[root@inst-centos ~]# showmount -e localhost
Export list for localhost:
/common 192.168.17.0/24
[root@inst-centos ~]#
```

Figure 105.1

```
192.168.0.3 - PuTTY
root@debian:~# showmount -e 192.168.0.10
Export list for 192.168.0.10:
/common 192.168.0.0/24
root@debian:~#
```

Figure 105.2

In the first example, I used showmount to display the exported (shared) files on my local NFS server.

In the second example, I used showmount to display the exported (shared) files on a remote NFS server.

S

showmount

@soundtraining

showmount: shows mount
information for an NFS server

30 minutes ago

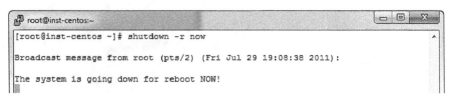

Figure 106.1

Commonly used options include -h (halt) and -r (restart). You must specify the time of the shutdown.

S

shutdown

@soundtraining

shutdown: as the name implies, shuts down the Linux system in a secure way.

2 minutes ago

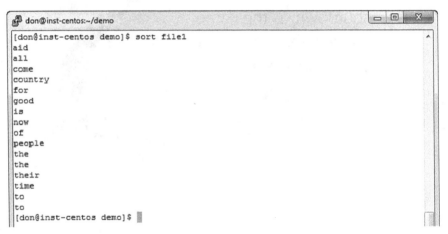

Figure 107.1

Figure 107.2

This is pretty self-explanatory.

S

sort

@soundtraining

sort: sorts line of text files

2 days ago

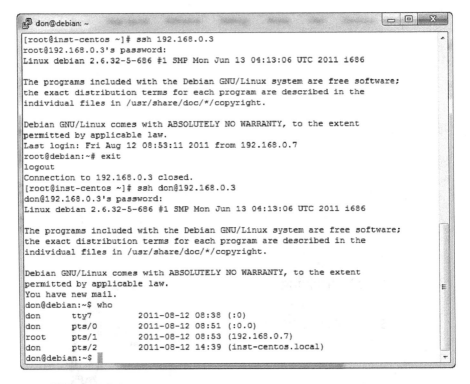

```
[root@inst-centos ~]# ssh 192.168.0.3
root@192.168.0.3's password:
Linux debian 2.6.32-5-686 #1 SMP Mon Jun 13 04:13:06 UTC 2011 i686

The programs included with the Debian GNU/Linux system are free software;
the exact distribution terms for each program are described in the
individual files in /usr/share/doc/*/copyright.

Debian GNU/Linux comes with ABSOLUTELY NO WARRANTY, to the extent
permitted by applicable law.
Last login: Fri Aug 12 08:53:11 2011 from 192.168.0.7
root@debian:~# exit
logout
Connection to 192.168.0.3 closed.
[root@inst-centos ~]# ssh don@192.168.0.3
don@192.168.0.3's password:
Linux debian 2.6.32-5-686 #1 SMP Mon Jun 13 04:13:06 UTC 2011 i686

The programs included with the Debian GNU/Linux system are free software;
the exact distribution terms for each program are described in the
individual files in /usr/share/doc/*/copyright.

Debian GNU/Linux comes with ABSOLUTELY NO WARRANTY, to the extent
permitted by applicable law.
You have new mail.
don@debian:~$ who
don      tty7        2011-08-12 08:38 (:0)
don      pts/0       2011-08-12 08:51 (:0.0)
root     pts/1       2011-08-12 08:53 (192.168.0.7)
don      pts/2       2011-08-12 14:39 (inst-centos.local)
don@debian:~$
```

Figure 108.1

The ssh program allows you to encrypt remote login sessions. In the first example, I logged in to the remote system (192.168.0.3) as the current local user (root). In the second example, I logged in as remote user don (don@192.168.0.3).

I also used the "who" command to display all login sessions on the remote system.

See also "who".

S

ssh

@soundtraining

ssh: is the OpenSSH client program which allows secure remote login

1 day ago

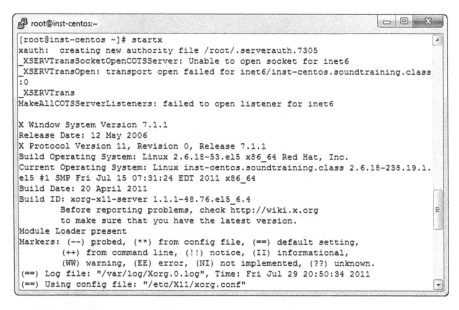

Figure 109.1

On most systems, you can kill X windows with the keyboard combination of Ctrl+Alt+Backspace.

S

startx

@soundtraining

startx: initializes X windows

54 minutes ago

```
[don@inst-centos demo]$ stat file1
  File: `file1'
  Size: 72              Blocks: 8          IO Block: 4096   regular file
Device: fd00h/64768d    Inode: 2047714     Links: 1
Access: (0664/-rw-rw-r--)  Uid: (  500/     don)   Gid: (  500/     don)
Access: 2011-07-29 20:53:10.000000000 -0700
Modify: 2011-07-29 20:40:54.000000000 -0700
Change: 2011-07-29 20:40:54.000000000 -0700
[don@inst-centos demo]$
```

Figure 110.1

```
[root@inst-centos ~]# stat /dev/sdb1
  File: `/dev/sdb1'
  Size: 0               Blocks: 0          IO Block: 4096   block special file
Device: 11h/17d Inode: 1229        Links: 1     Device type: 8,11
Access: (0640/brw-r-----)  Uid: (    0/    root)   Gid: (    6/    disk)
Access: 2011-08-12 11:50:45.685815577 -0700
Modify: 2011-08-12 11:50:45.869807198 -0700
Change: 2011-08-12 11:50:45.869807198 -0700
[root@inst-centos ~]#
```

Figure 110.2

In the first example, I ran stat on an individual file (file1). In the second, I ran it on a file system (/dev/sdb1).

See also "file".

S

stat

@soundtraining

stat: displays file or file system status

1 day ago

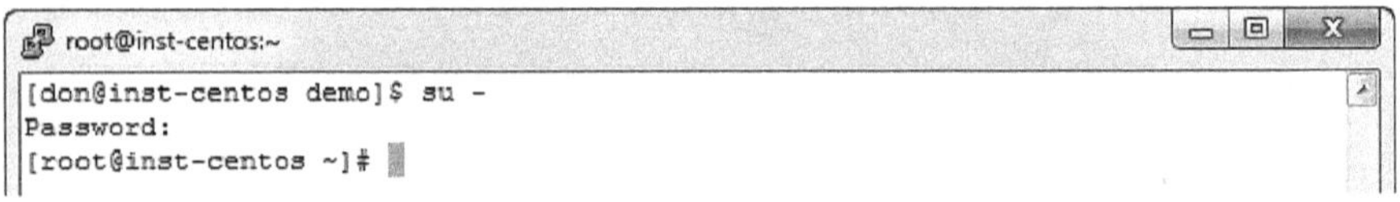

```
root@inst-centos:~                                    [_][□][X]
[don@inst-centos demo]$ su -
Password:
[root@inst-centos ~]#
```

Figure 111.1

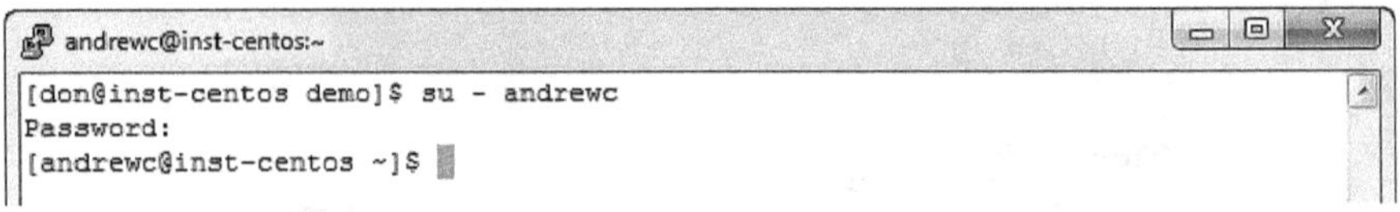

```
andrewc@inst-centos:~                                 [_][□][X]
[don@inst-centos demo]$ su - andrewc
Password:
[andrewc@inst-centos ~]$
```

Figure 111.2

It's more secure to perform administrative tasks with "sudo" than with "su", but it's easier with "su". I find that, when I need to spend a lot of time managing a server, I tend to use "su". When I just need to execute a few commands, I use "sudo". Just be careful when you "su" to root.

See also "sudo".

S

su

@soundtraining

su: is the switch user program. When used with the "-" option, it includes the new user's profile

30 seconds ago

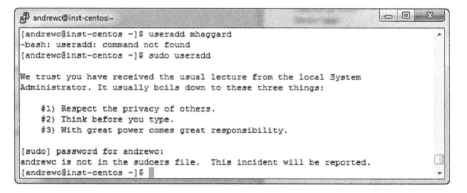

Figure 112.1

In this example, user andrewc, who is not authorized in the sudoers file to perform administrative tasks, attempted to add a user. He was not permitted to perform the task and the incident was written to the security log (/var/log/secure):

(continued on page 244)

S

sudo

@soundtraining

sudo: allows programs to run with another user, often root

22 seconds ago

```
Aug 12 15:14:45 inst-centos sudo:  andrewc : user NOT in sudoers ; TTY=pts/3 ; P
WD=/home/andrewc ; USER=root ; COMMAND=useradd
```

Figure 112.2

Notice that /var/log/secure shows the time the incident occurred, the user in question (andrewc), the fact that the user is not in sudoers, the terminal from which the attempt occurred (TTY=pts/3), and the command he attempted to execute (useradd).

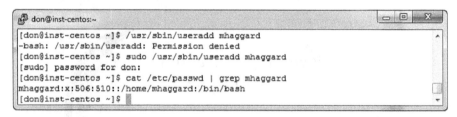

```
don@inst-centos:~
[don@inst-centos ~]$ /usr/sbin/useradd mhaggard
-bash: /usr/sbin/useradd: Permission denied
[don@inst-centos ~]$ sudo /usr/sbin/useradd mhaggard
[sudo] password for don:
[don@inst-centos ~]$ cat /etc/passwd | grep mhaggard
mhaggard:x:506:510::/home/mhaggard:/bin/bash
[don@inst-centos ~]$
```

Figure 112.3

In this example, I attempted to run the useradd command as myself. Permission was denied. I then used "sudo" with the command and, after supplying my password, the command was executed. The reason it worked is because I am listed in the computer's sudoers file.

See also "visudo".

S

sudo

@soundtraining

sudo: allows programs to run with another user, often root

22 seconds ago

```
don@inst-centos:~/demo
[don@inst-centos demo]$ tail /etc/httpd/conf/httpd.conf
    JkMount /*.cfres ajp13
    JkMount /*.cfm/* ajp13
    JkMount /*.cfml/* ajp13
    # Flex Gateway Mappings
    # JkMount /flex2gateway/* ajp13
    # JkMount /flashservices/gateway/* ajp13
    # JkMount /messagebroker/* ajp13
    JkMountCopy all
    JkLogFile /var/log/httpd/mod_jk.log
</IfModule>
[don@inst-centos demo]$ tail -n 12 /etc/httpd/conf/httpd.conf
    JkMount /*.jsp ajp13
    JkMount /*.cfchart ajp13
    JkMount /*.cfres ajp13
    JkMount /*.cfm/* ajp13
    JkMount /*.cfml/* ajp13
    # Flex Gateway Mappings
    # JkMount /flex2gateway/* ajp13
    # JkMount /flashservices/gateway/* ajp13
    # JkMount /messagebroker/* ajp13
    JkMountCopy all
    JkLogFile /var/log/httpd/mod_jk.log
</IfModule>
[don@inst-centos demo]$
```

Figure 113.1

The "tail" command is very useful when you just want to look at the bottom of a file. For example, when you want to see the most recently added users in /etc/passwd.

See also "head".

t

tail

@soundtraining

tail: outputs the last part of a file

2 seconds ago

```
don@inst-centos:~/demo                                          — ▢ ✕
[don@inst-centos demo]$ tar cvfj files.tgz file*
file1
file2
file3
[don@inst-centos demo]$ ls
file1  file2  file3  files.tgz
[don@inst-centos demo]$ rm file1 file2 file3
[don@inst-centos demo]$ ls
files.tgz
[don@inst-centos demo]$ tar xvf files.tgz
file1
file2
file3
[don@inst-centos demo]$ ls
file1  file2  file3  files.tgz
[don@inst-centos demo]$ 
```

Figure 114.1

Using the "tar" (short for "tape archive") command creates what
is called a "tarball". Combining multiple files into a single file
makes it easier to move files across networks, especially when you
compress them as well.

t

tar

@soundtraining

tar: is an archiving utility. You
can use it to combine multiple
files into a single file and, if
desired, compress the new file

13 minutes ago

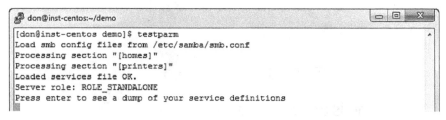

Figure 115.1

This is a great way to test smb.conf for syntax errors before attempting to restart your SAMBA server.

t

testparm

@soundtraining

testparm: checks the smb.conf
file for correctness

1 day ago

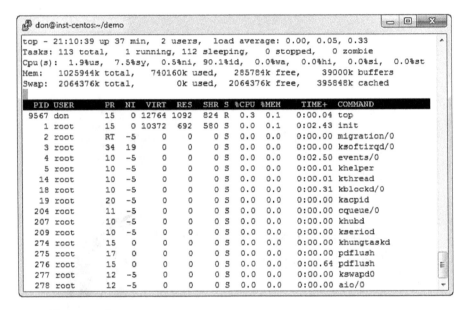

Figure 116.1

Unlike "ps aux" which provides a snapshot of all running processes, "top" provides a continuously updating (every three seconds, by default) view of your system's running processes.

t

top

@soundtraining

top: provides a dynamic real-time view of a running system

1 day ago

```
don@inst-centos:~/demo                                          _  □  X
[don@inst-centos demo]$ ls -l
total 16
-rw-rw-r-- 1 don don  72 Jul 29 20:40 file1
-rw-rw-r-- 1 don don  58 Jul 29 21:04 file2
-rw-rw-r-- 1 don don  55 Jul 29 21:04 file3
-rw-rw-r-- 1 don don 289 Jul 29 21:05 files.tgz
[don@inst-centos demo]$ touch -t 1408281130 file1
[don@inst-centos demo]$ ls -l
total 16
-rw-rw-r-- 1 don don  72 Aug 28  2014 file1
-rw-rw-r-- 1 don don  58 Jul 29 21:04 file2
-rw-rw-r-- 1 don don  55 Jul 29 21:04 file3
-rw-rw-r-- 1 don don 289 Jul 29 21:05 files.tgz
[don@inst-centos demo]$ touch file4
[don@inst-centos demo]$ ls -l
total 16
-rw-rw-r-- 1 don don  72 Aug 28  2014 file1
-rw-rw-r-- 1 don don  58 Jul 29 21:04 file2
-rw-rw-r-- 1 don don  55 Jul 29 21:04 file3
-rw-rw-r-- 1 don don   0 Jul 29 21:13 file4
-rw-rw-r-- 1 don don 289 Jul 29 21:05 files.tgz
[don@inst-centos demo]$
```

Figure 117.1

While "touch" is officially designed to change timestamps, lots of us use it to create blank files for testing purposes.

t

touch

@soundtraining

touch: changes file timestamps. It can also be used to create new, blank files

2 days ago

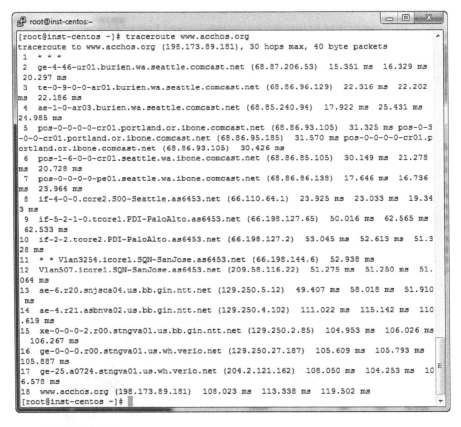

Figure 118.1

In this example, I started a traceroute from my home in Seattle to www.acchos.org, the website of the Atlantic City Convention Hall Organ Society (dedicated to the preservation and maintenance of the world's largest pipe organ ... you really should check out their website). As you can see, it went through 18 hops to get from Seattle to its destination on a Verio server.

See also "ping" and "mtr".

traceroute

@soundtraining

traceroute: displays the route followed by packets to a remote host, including router hops along the way

39 minutes ago

```
Ubuntu 10.04.2 LTS ubuntu tty1

ubuntu login: don
Password:
Last login: Fri Aug 12 16:26:57 PDT 2011 on tty2
Linux ubuntu 2.6.32-28-generic #55-Ubuntu SMP Mon Jan 10 23:42:43 UTC 2011 x86_6
4 GNU/Linux
Ubuntu 10.04.2 LTS

Welcome to Ubuntu!
 * Documentation:  https://help.ubuntu.com/
don@ubuntu:~$ tty
/dev/tty1
don@ubuntu:~$ _
```

Figure 119.1

I use the tty command most often when I have multiple console sessions going (from using alt+function keys) and I can't remember which one is which.

tty
@soundtraining

tty: displays the terminal pathname

27 minutes ago

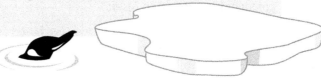

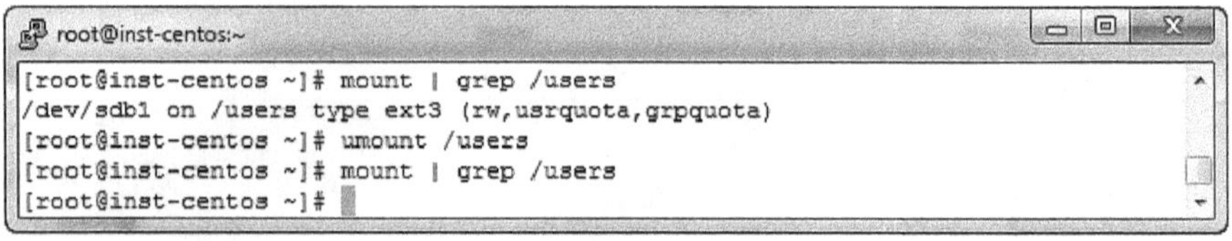

```
root@inst-centos:~
[root@inst-centos ~]# mount | grep /users
/dev/sdb1 on /users type ext3 (rw,usrquota,grpquota)
[root@inst-centos ~]# umount /users
[root@inst-centos ~]# mount | grep /users
[root@inst-centos ~]#
```

Figure 120.1

In the above example, I used "mount | grep /users" to show that /dev/sdb1 was mounted on the directory /users. Then, I used "umount /users" to unmount it which is shown when I repeated the command "mount | grep /users".

umount

@soundtraining

umount: detaches the specified filesystem from the local system's file hierarchy

9 minutes ago

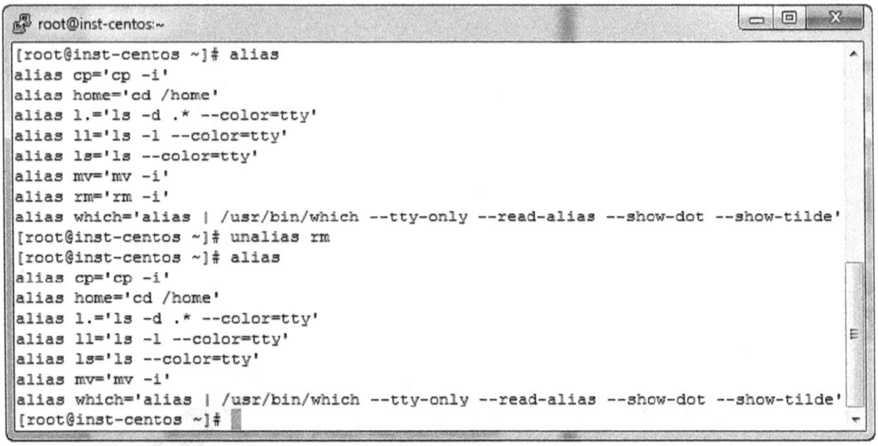

Figure 121.1

In this example, I first used the alias command to display all existing aliases. Then, I removed the alias "alias rm='rm –i' with the unalias command. Finally, I used again used the alias command to demonstrate that the alias "rm" had indeed been removed.

See also "alias".

u

unalias

@soundtraining

unalias: disables a previously-configured alias, including those configured by a user or global profile

17 minutes ago

```
don@inst-centos:~/demo                                    _  □  X
[don@inst-centos demo]$ uname -o
GNU/Linux
[don@inst-centos demo]$ uname -v
#1 SMP Fri Jul 15 07:31:24 EDT 2011
[don@inst-centos demo]$ uname -i
x86_64
[don@inst-centos demo]$ uname -r
2.6.18-238.19.1.el5
[don@inst-centos demo]$ uname -a
Linux inst-centos.soundtraining.class 2.6.18-238.19.1.el5 #1 SMP Fri Jul 15 07:3
1:24 EDT 2011 x86_64 x86_64 x86_64 GNU/Linux
[don@inst-centos demo]$
```

Figure 122.1

The uname command is very helpful when you need to
understand things about your system like processor architecture
or kernel version.

u

uname

@soundtraining

uname: displays various system information including kernel information and processor information

1 day ago

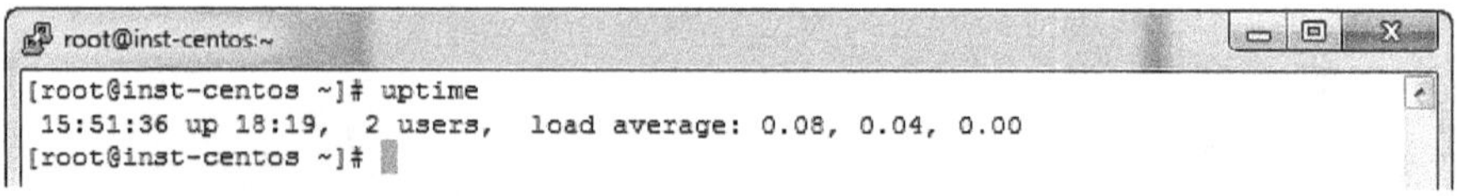

```
root@inst-centos:~
[root@inst-centos ~]# uptime
 15:51:36 up 18:19,  2 users,  load average: 0.08, 0.04, 0.00
[root@inst-centos ~]#
```

Figure 123.1

uptime gives a one line display of the following information. The current time, how long the system has been running, how many users are currently logged on, and the system load averages for the past 1, 5, and 15 minutes. It's the same information that is displayed in the header of the "w" command.

See also the "w" command.

u

uptime

@soundtraining

uptime: tells how long the system has been running, number of logged on users, and system load averages

26 minutes ago

```
root@centos6:~
[root@centos6 ~]# useradd donc
[root@centos6 ~]# useradd -c "Andrew Crawley" -e 2012-09-01 andrewc
[root@centos6 ~]# 
```

Figure 124.1

In this example, I first added a user named donc with no options.
I then added a user named andrewc, but included a comment
(-c) of his full name, and an account expiration date (-e) of
September 1, 2012. As you can imagine, there are many other
options. Check the info page for a complete list.

See also 'adduser", "usermod", and "userdel".

U

useradd

@soundtraining

useradd: creates a new user
or updates default new
user information

2 days ago

```
root@inst-centos:~

[root@inst-centos ~]# cat /etc/passwd | grep lgaga
lgaga:x:501:502:Lady Gaga:/home/lgaga:/bin/bash
[root@inst-centos ~]# ls /home
andrewc  don  donc  jonc  lgaga  user01
[root@inst-centos ~]# userdel -r lgaga
[root@inst-centos ~]# cat /etc/passwd | grep lgaga
[root@inst-centos ~]# ls /home
andrewc  don  donc  jonc  user01
[root@inst-centos ~]#
```

Figure 125.1

While userdel is a pretty straightforward command, this example
shows its use with the "-r" option which, in addition to deleting the
user account, also deletes that user's home directory and all
its contents.

See also "adduser", "useradd", and "usermod".

u

userdel

@soundtraining

userdel: deletes a user account

47 minutes ago

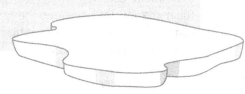

```
root@inst-centos:~
[root@inst-centos ~]# cat /etc/passwd | grep jonc
jonc:x:504:508::/home/jonc:/bin/bash
[root@inst-centos ~]# usermod -c "Jon Y. Crawley" jonc
[root@inst-centos ~]# cat /etc/passwd | grep jonc
jonc:x:504:508:Jon Y. Crawley:/home/jonc:/bin/bash
[root@inst-centos ~]#
```

Figure 126.1

The usermod command does what the name suggests: it modifies
a user account. In this example, I used the "cat /etc/passwd | grep
jonc" command to display jonc's user account in /etc/passwd. As
you can see, his full name is not included as part of his account
information. I then used usermod with the "-c" switch to add
a comment. Usually, comments in user accounts are the user's
full name. Finally, I repeated the "cat /etc/passwd | grep jonc"
command to demonstrate that his full name was added as a
comment to his user account.

See also "adduser", "useradd", and "userdel".

usermod

@soundtraining

**usermod: modifies
a user account**

22 seconds ago

```
root@inst-centos:~
[root@inst-centos ~]# users
andrewc andrewc don jonc root
[root@inst-centos ~]#
```

Figure 127.1

This simple, straightforward utility shows a list of currently logged on users. The only options are "version" and "help".

Notice that it displays sessions, thus user andrewc is listed twice because he's logged on twice.

See also "who".

u

users

@soundtraining

users: displays the usernames
of currently logged-on users

1 day ago

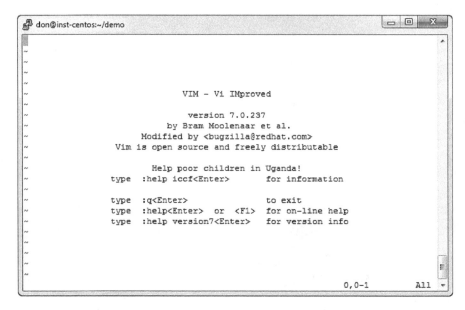

```
don@inst-centos:~/demo

~
~
~
~
~                        VIM - Vi IMproved
~
~                       version 7.0.237
~                    by Bram Moolenaar et al.
~                Modified by <bugzilla@redhat.com>
~         Vim is open source and freely distributable
~
~                 Help poor children in Uganda!
~         type   :help iccf<Enter>        for information
~
~         type   :q<Enter>                to exit
~         type   :help<Enter>  or  <F1>   for on-line help
~         type   :help version7<Enter>    for version info
~
~
~
~
~
~
~
                                                   0,0-1            All
```

Figure 128.1

Sometimes, I read or hear Linux admins (especially beginners) complaining about vi or its successor vim. They say its command structure is arcane (it is) and that it's old (it is). They wonder about the value of learning it, but when you work in the command-line environment as most Linux/Unix server admins do, vi(m) becomes an incredibly valuable tool. It's not really very difficult to gain proficiency in its use. Once you gain that proficiency, you can work incredibly fast. Perhaps the most important reason for learning vi(m), however, is that it's ubiquitous. Nearly every Linux/Unix server has it installed. If you need to modify a configuration file while working in a command-line environment, you can almost always count on having vi(m) available.

See also "vimtutor".

V

vi

@soundtraining

vi: starts the vi text editor
(usually aliased to vim)

12 minutes ago

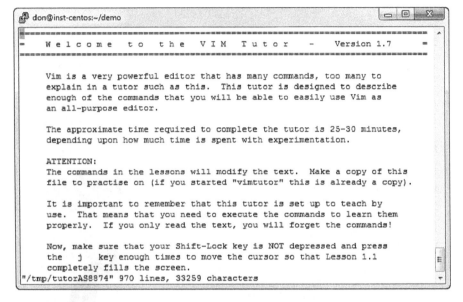

Figure 129.1

On the vi page, I ranted about the value of learning the vi(m) text editor. This tool, included in every vi installation I've ever seen, will help you become proficient with vi(m) in about 30 minutes. Really.

You can also download the free vim cheat sheet from www. soundtraining.net and tape it to the side of your monitor.

See also "vi".

vimtutor

@soundtraining

vimtutor: starts a vim tutorial

44 minutes ago

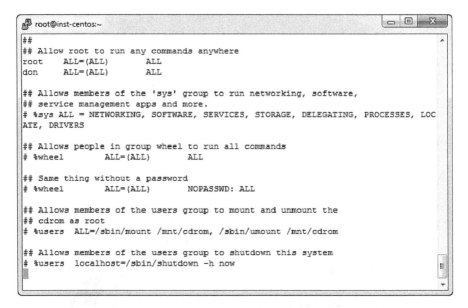

```
root@inst-centos:~
##
## Allow root to run any commands anywhere
root    ALL=(ALL)       ALL
don     ALL=(ALL)       ALL

## Allows members of the 'sys' group to run networking, software,
## service management apps and more.
# %sys ALL = NETWORKING, SOFTWARE, SERVICES, STORAGE, DELEGATING, PROCESSES, LOC
ATE, DRIVERS

## Allows people in group wheel to run all commands
# %wheel        ALL=(ALL)       ALL

## Same thing without a password
# %wheel        ALL=(ALL)       NOPASSWD: ALL

## Allows members of the users group to mount and unmount the
## cdrom as root
# %users  ALL=/sbin/mount /mnt/cdrom, /sbin/umount /mnt/cdrom

## Allows members of the users group to shutdown this system
# %users  localhost=/sbin/shutdown -h now
```

Figure 130.1

This is how you add users to sudoers to grant them elevated privileges. In this example, I added myself with the line:

```
don    ALL=(ALL)       ALL
```

That means that the user don has permission to run from any terminal (the first ALL), acting as any user (the second ALL), and execute any command (the third ALL).

See also "sudo".

V

visudo

@soundtraining

visudo: opens the sudoers file for editing

36 seconds ago

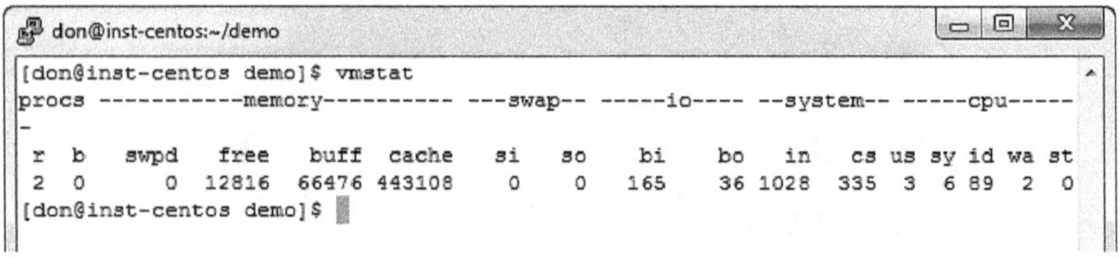

Figure 131.1

The vmstat command provides a summary overview of hardware performance.

V

vmstat

@soundtraining

vmstat: reports information about processes, memory, paging, block IO, traps, and cpu activity

1 day ago

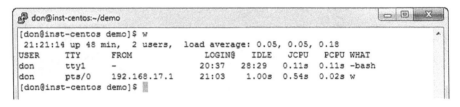

Figure 132.1

The one-letter command "w" combines "uptime" and "who" in a single command.

See also "uptime" and "who".

W

@soundtraining

w: shows who is logged on, what they are doing, and system load information

54 minutes ago

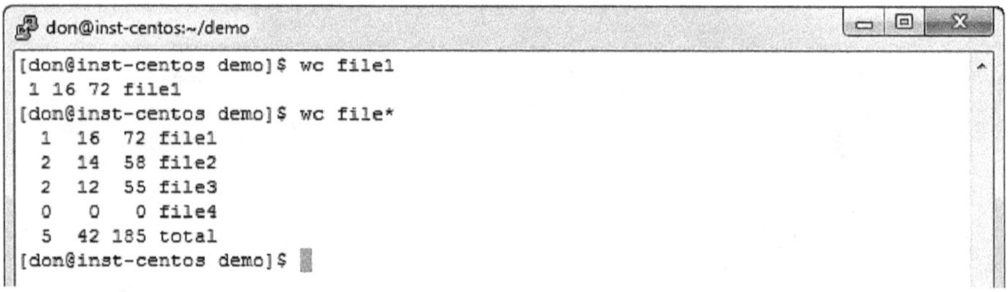

```
don@inst-centos:~/demo                                    □  ▣  ✕
[don@inst-centos demo]$ wc file1
 1 16 72 file1
[don@inst-centos demo]$ wc file*
  1   16   72 file1
  2   14   58 file2
  2   12   55 file3
  0    0    0 file4
  5   42  185 total
[don@inst-centos demo]$
```

Figure 133.1

Not necessarily a system administrator's tool, but helpful nonetheless, wc gives you the word count(s) for a file or files.

W

WC
@soundtraining

wc: displays newline, word, and byte counts for each FILE, and a total line if more than one file is specified

11 minutes ago

```
root@inst-centos:~/downloads                                        _ □ X
[root@inst-centos downloads]# wget ftp://ftp.isc.org/isc/bind9/9.7.4/bind-9.7.4. ▲
tar.gz
--2011-08-08 12:20:30--  ftp://ftp.isc.org/isc/bind9/9.7.4/bind-9.7.4.tar.gz
           => `bind-9.7.4.tar.gz'
Resolving ftp.isc.org... 204.152.184.110, 2001:4f8:0:2::18
Connecting to ftp.isc.org|204.152.184.110|:21... connected.
Logging in as anonymous ... Logged in!
==> SYST ... done.    ==> PWD ... done.
==> TYPE I ... done.   ==> CWD /isc/bind9/9.7.4 ... done.
==> SIZE bind-9.7.4.tar.gz ... 8316839
==> PASV ... done.    ==> RETR bind-9.7.4.tar.gz ... done.
Length: 8316839 (7.9M)

100%[===================================>] 8,316,839    768K/s   in 11s

2011-08-08 12:20:41 (772 KB/s) - `bind-9.7.4.tar.gz' saved [8316839]

[root@inst-centos downloads]# ls
bind-9.7.4.tar.gz
[root@inst-centos downloads]# ▮
```

Figure 134.1

This is very helpful utility for downloading files from the Internet.
I use it most frequently to download installation tarballs. There is a
version for Windows, too.

W

wget

@soundtraining

wget: is a non-interactive Web downloader. It is useful for retrieving files from websites while working in a command-line environment

2 days ago

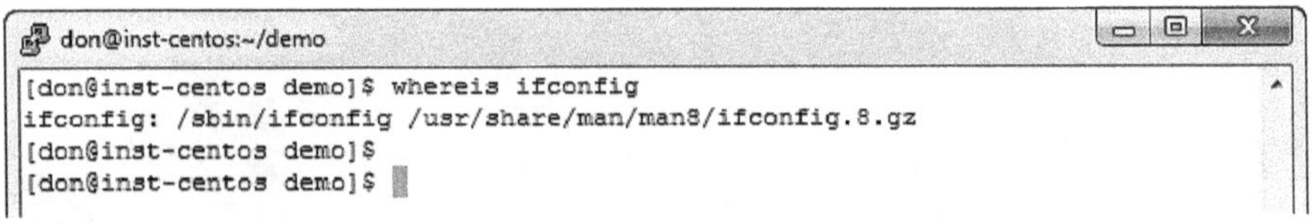

```
don@inst-centos:~/demo                                          ☐ ▣ ✕
[don@inst-centos demo]$ whereis ifconfig
ifconfig: /sbin/ifconfig /usr/share/man/man8/ifconfig.8.gz
[don@inst-centos demo]$
[don@inst-centos demo]$ ▌
```

Figure 135.1

The whereis command is useful for finding files related to a particular command.

See also "which", "find", and "locate".

W

whereis

@soundtraining

whereis: locates the binary, source, and manual page files for a command

10 minutes ago

```
root@inst-centos:~                                          _  □  X
[root@inst-centos ~]# which ifconfig
/sbin/ifconfig
[root@inst-centos ~]# which named
/usr/sbin/named
[root@inst-centos ~]# which httpd
/usr/sbin/httpd
[root@inst-centos ~]# which ls
alias ls='ls --color=tty'
        /bin/ls
[root@inst-centos ~]#
```

Figure 136.1

The "which" command is a down-and-dirty tool for displaying the full path of commands. It is especially useful when you move from one distro to another and you're not sure where things are.

See also "whereis", "find", and "locate".

W

which

@soundtraining

which: displays the full path
of commands

1 day ago

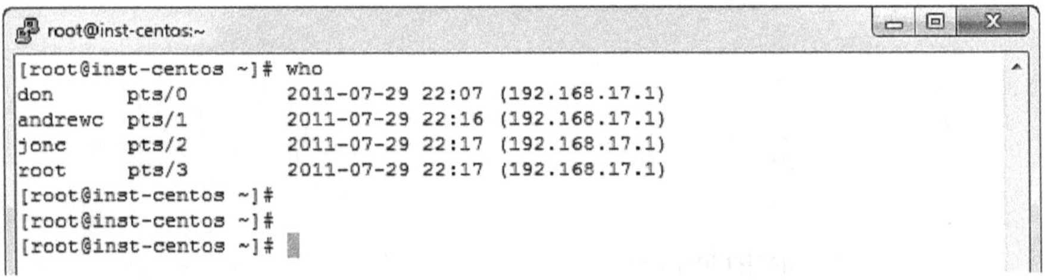

```
root@inst-centos:~
[root@inst-centos ~]# who
don        pts/0       2011-07-29 22:07 (192.168.17.1)
andrewc    pts/1       2011-07-29 22:16 (192.168.17.1)
jonc       pts/2       2011-07-29 22:17 (192.168.17.1)
root       pts/3       2011-07-29 22:17 (192.168.17.1)
[root@inst-centos ~]#
[root@inst-centos ~]#
[root@inst-centos ~]#
```

Figure 137.1

There's not much to say here. It shows the users who are currently logged on and the terminals they're using.

See also "w".

W

who

@soundtraining

who: shows who is logged on to the system

2 seconds ago

Figure 138.1

We've all heard the jokes about not knowing who we are, but this is truly useful when you've got lots of terminals open. Frankly, I usually include the current user name in the prompt, so I don't have to worry about it. Still, this command is handy when you're working on someone else's system.

W

whoami

@soundtraining

whoami: prints the username
of the currently logged on user

30 seconds ago

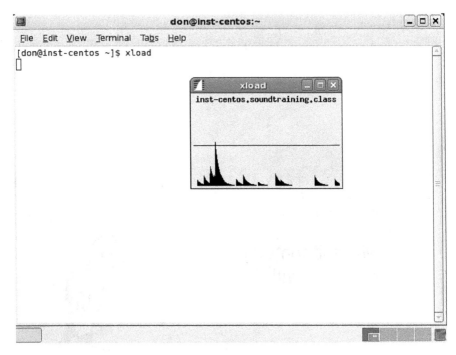

Figure 139.1

Honestly, I don't find this command very useful, but the histogram is kind of cool.

X

xload

@soundtraining

xload: shows a periodically updated histogram of the system load average

1 day ago

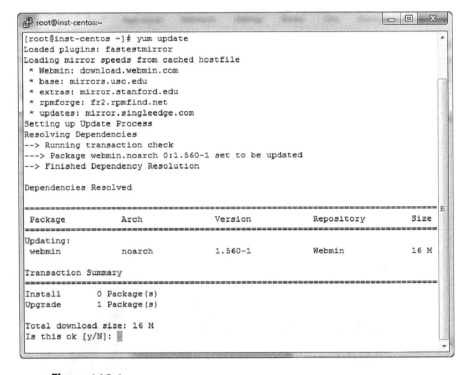

Figure 140.1

The yum utility (Yellow Dog Updated Modified) is the most common way to manage packages (applications) on RedHat-based systems, but other distros also support the use of yum.

Use 'yum search [package name" to see what's available, "yum install [package name]" to install new software, or "yum update [package name] to update software.

See also "apt-get" and "aptitude".

y

yum
@soundtraining

yum: is a package management system to search for, install, update, and remove packages (applications)

22 minutes ago

```
don@inst-centos:~/demo
[don@inst-centos demo]$ zip files.zip file*
  adding: file1 (deflated 15%)
  adding: file2 (deflated 3%)
  adding: file3 (stored 0%)
  adding: file4 (stored 0%)
  adding: file5 (deflated 65%)
[don@inst-centos demo]$ ls -l
total 64
-rw-rw-r-- 1 don don    72 Jul 29 22:10 file1
-rw-rw-r-- 1 don don    58 Jul 29 21:04 file2
-rw-rw-r-- 1 don don    55 Jul 29 21:04 file3
-rw-rw-r-- 1 don don     0 Jul 29 21:13 file4
-rw-r--r-- 1 don don 34278 Jul 29 23:14 file5
-rw-rw-r-- 1 don don 12707 Jul 29 23:19 files.zip
[don@inst-centos demo]$
```

Figure 141.1

My observation is that most of us use either gzip or bzip2 for file compression on Linux systems. Still, it's handy to have the zip utility for compatibility with systems running the Windows operating system.

See also "bzip2" and "gzip".

Z

zip

@soundtraining

zip: compresses and packages files.
It is compatible with PKZIP for
MSDOS systems

1 second ago

Appendix: Helpful Websites

Linux Training:

www.soundtraining.net

Miscellaneous Linux Websites:

www.li.org

www.linux.org

www.linuxquestions.org

www.linuxfordevices.com

www.tldp.org

www.samba.org

www.kernel.org

www.sendmail.org

www.ietf.org

www.rfc-editor.org

www.linuxtoday.com

www.freshmeat.net

www.gnome.org

www.kde.org

www.openoffice.org

www.apache.org

www.postfix.org

www.putty.org

www.realvnc.com

www.tightvnc.com

www.uvnc.com

www.rpm.org

www.linuxfoundation.org

www.zoneedit.com

www.openldap.org

www.netfilter.org

www.shelldorado.com

Problems with sound cards, try this:

www.alsa-project.org

For Linux printer drivers:

www.linuxprinting.org

www.linuxprinting.org/foomatic.html

http://gimp-print.sourceforge.net

Some Linux Distributions:

www.redhat.com

www.suse.com

www.slackware.com

www.knoppix.net

www.debian.org

www.yellowdoglinux.com

www.gentoo.org

www.ubuntu.com

www.centos.org

www.distrowatch.com

Other books by Don R. Crawley ...

The Accidental Administrator®: Linux Server Step-by-Step Configuration Guide

Packed with 44 easy-to-follow hands-on exercises plus numerous command examples and screen captures to help you build a working Linux server configuration from scratch. It's the most straight-forward approach to learning how to configure a CentOS/Red Hat/Fedora Linux server (the book is based on version 5.4 and 5.5), filled with practical tips and secrets learned from years of teaching, consulting, and administering Linux servers. There is no time wasted on boring theory. The essentials are covered in chapters on installing, administering, user management, file systems and directory management, networking, package management, automated task scheduling, network services, Samba, NFS, disk quotas, mail servers, Web and FTP servers, desktop sharing, printing, security, routing, performance monitoring, management tools, and more.

ISBN: 978-1-45368-992-9

The Accidental Administrator®: Cisco ASA Step-by-Step Configuration Guide

Packed with 56 easy-to-follow hands-on exercises to help you build a working firewall configuration from scratch. It's the most straight-forward approach to learning how to configure the Cisco ASA Security Appliance, filled with practical tips and secrets learned from years of teaching and consulting on the ASA. The essentials are covered in chapters on installing, backups and restores, remote administration, VPNs, DMZs, usernames, transparent mode, static NAT, port address translation, access lists, DHCP, password recovery, logon banners, AAA (authentication, authorization, and accounting), filtering content, and more.

ISBN: 978-1-44959-662-0

The Compassionate Geek: Mastering Customer Service for I.T. Professionals

Co-authored with Paul R. Senness, now in its second edition, is a customer service book written especially for today's overworked I.T. staff! Filled with practical tips, best practices, and real-world techniques, *The Compassionate Geek* is a quick read with equally fast results. Learn how to speak to the different generations at work, how to use emotional intelligence to manage your own emotions and influence the emotions of others, how to say "no" without alienating the end-user, what to do when the customer (user) is wrong, how to cope with the stress of the job, and more! All of the information is presented in a straightforward style that you can understand and use right away. There's nothing "foo-foo", just down-to-earth tips and best practices learned from years of working with I.T. pros and end-users.

ISBN: 978-0-98366-070-5

Also available as an e-book in various formats!
(Please check availability with your favorite e-book retailer.)

The Accidental Administrator®: Cisco Router Step-by-Step Configuration Guide

Tweeting Linux: 140 Linux Commands Explained in 140 Characters or Less

The Accidental Administrator®: Linux Server Step-by-Step Configuration Guide

The Compassionate Geek: Mastering Customer Service for I.T. Professionals

WWW.SOUNDTRAINING.NET/BOOKSTORE

Onsite Training
Makes Sense!

Learning solutions that come right to your door!

One- and two-day seminars and workshops for I.T. professionals

soundtraining.net
accelerated i.t. training

Call (206) 988-5858 • soundtraining.net/onsite • Email: onsite@soundtraining.net